GOODS OF THE MIND, LLC

Competitive Mathematics Series

for

Gifted Students in Grades 5 and 6

PRACTICE OPERATIONS
AND ALGEBRA

Cleo Borac, M. Sc.
Silviu Borac, Ph. D.

This edition published in 2016 in the United States of America.

Editing and proofreading: David Borac, B.Mus.
Technical support: Andrei T. Borac, B.A., PBK

Send all inquiries to:

Goods of the Mind, LLC
1138 Grand Teton Dr.
Pacifica
CA, 94044

Competitive Mathematics Series for Gifted Students
Level III (Grades 5 and 6)
Operations and Algebra

Contents

Foreword

The goal of these booklets is to provide a problem solving training ground starting from the earliest years of a student's mathematical development.

In our experience, we have found that teaching how to solve problems should focus not only on finding correct answers but also on finding better solution strategies. While the correct answer to a problem can typically be obtained in several different ways, not all these ways are equally useful for learning how to solve problems.

The most basic strategy is *brute force*. For example, if a problem asks for the number of ways Lila and Dina can sit on a bench, it is easy to write down all the possibilities: Dina, Lila and Lila, Dina. We arrive at this solution by performing all the possible actions allowed by the problem, leaving nothing to the imagination. For this last reason, this approach is called brute force.

Obviously, if we had to figure out the number of ways 30 people could stand in a line, then brute force would not be as practical, as it would take a prohibitively long time to apply.

Using brute force to obtain the correct answer for a simpler problem is not necessarily a useful learning experience for solving a similar problem that is more complex. Moreover, solving problems in a quantitative manner, assuming that the student can transfer simple strategies to similar but more complex problems, is not an efficient way of learning problem solving.

From this simple example, we see that the goal of *practicing* problem solving is different from the goal of problem solving. While the goal of problem solving is to obtain a correct answer, the goal of practicing problem solving is to acquire the ability to develop strategies, generate ideas, and combine approaches that are powerful enough to solve the problem at hand as well as future similar problems.

While brute force is not a useless strategy, it is not a key that opens every door. Nevertheless, there are problems where brute force can be a useful tool. For instance, brute force can be used as a first step in solving a complex problem: a smaller scale example can be approached using brute force to help the problem solver understand the mechanics of the problem and generate ideas for solving the larger case.

All too often, we encounter students who can quickly solve simple problems by applying brute force and who become frustrated when the solving methods they have been employing successfully for years become inefficient once problems increase in complexity. Often, neither the student nor the parent has a clear understanding of why the student has stagnated at a certain level. When the only arrows in the quiver are guess-and-check and brute force, the ability to take down larger game is limited.

Our series of books aims to address this tendency to continue on the beaten path - which usually generates so much praise for the gifted student in the early years of schooling - by offering a challenging set of questions meant to build up an understanding of the problem solving process. Solving problems should never be easy! To be useful, to represent actual training, problem solving should be challenging. There should always be a sense of difficulty, otherwise there is no elation upon finding the solution.

Indeed, practicing problem solving is important and useful only as a means of learning how to develop better strategies. We must constantly learn and invent new strategies while questioning the limitations of the strategies we are using. Obtaining the correct answer is only the natural outcome of having applied a strategy that worked for a particular problem in the time available to solve it. Obtaining the wrong answer is not necessarily a bad outcome; it provides insight into the fallacies of the method used or into the errors of execution that may have occured. As long as students manifest an interest in figuring out strategies, the process of problem solving should be rewarding in itself.

Sitting and thinking in a focused manner is difficult to train, particularly since the modern lifestyle is not conducive to adopting open-ended activities. This is why we would like to encourage parents to pull back from a quantitative approach to mathematical education based on repetition, number of completed pages, and the number of correct answers. Instead, open up the time boundaries that are dedicated to math, adopt math as a game played in the family, initiate a math dialogue, and let the student take his or her time to think up clever solutions.

Figuring out strategies is much more of a game than the mechanical repetition of stepwise problem solving recipes that textbooks so profusely provide, in order to "make math easy." Mathematics is not meant to be easy; it is meant to be interesting.

Solving a problem in different ways is a good way of comparing the merits of each method - another reason for not making the correct answer the primary goal of the activity. Which method is more labor intensive, takes more time or is more prone to execution errors? These are questions that must be part of the problem solving process.

In the end, it is not the quantity of problems solved, the level of theory absorbed, or the number of solutions offered in ready-made form by so many courses and camps, but the willingness to ask questions, understand and explore limitations, and derive new information from scratch, that are the cornerstones of a sound training for problem solvers.

These booklets are not a complete guide to the problem solving universe, but they are meant to help parents and educators work in the direction that, aside from being the most efficient, is the more interesting and rewarding one.

The series is designed for mathematically gifted students. Each book addresses an age range as some students will be ready for this content earlier, others later. If a topic seems too difficult, simply try it again in a couple of months.

ARITHMETIC SEQUENCES

Sequences are sets of numbers that can be derived by using a rule. The elements of a sequence are called *terms*. The position of a term within the sequence is calculated from the first term and is called a *rank*.

The current term of an **arithmetic sequence** can be generated by adding/subtracting a number to/from the previous term. This number is called the *common difference* of the sequence. All the terms of the sequence can be found if we know the first term and the common difference.

Rank	1st	2nd	3rd	4th	5th	6th
Sequence terms	9	14	19	24	29	34
Common difference		5	5	5	5	5

The **arithmetic mean** is the average of two numbers. In an arithmetic sequence, any term except the first and last ones is the arithmetic mean of its neighbors.

Consider an arithmetic sequence $\{a_j\}$ with common difference d and a_k a term of rank k. Then, the terms of rank $k-1$ (previous term) and $k+1$ (next term) are given by:

$$a_{k-1} = a_k - d$$
$$a_{k+1} = a_k + d$$

We notice that:

$$a_{k-1} + a_{k+1} = 2a_k$$

hence the term a_k is the arithmetic mean of its neighbors:

$$a_k = \frac{a_{k-1} + a_{k+1}}{2}$$

9

The sum of N consecutive integers starting at 1 is the most common example of arithmetic sequence. Can you figure out what the first term and the common difference are for this sequence?

Denote the sum with S:

$$S = 1 + 2 + 3 + \cdots + (N - 1) + N$$

Write the same sum again, reversing the terms:

$$S = N + (N - 1) + \cdots + 3 + 2 + 1$$

Add the two sums by grouping vertically. We obtain the double of S:

```
S  =   1    +   2   +  ···  +  (N-1)  +   N
S  =   N    + (N-1) +  ···  +    2    +   1
_____  +
2S  = (N+1) + (N+1) +  ···  +  (N+1)  + (N+1)
```

which is N times $(N + 1)$, and we divide by 2 to find S:

$$S = \frac{N(N + 1)}{2}$$

Notice that this process works regardless of whether N is even or odd.

Numbers of the form S are called *triangular numbers*.

The sum of 2k consecutive even integers starting at 2

Denote the sum with E:

$$E = 2 + 4 + 6 + \cdots + 2k$$

Since all the terms are multiples of 2, factor out a 2:

$$E = 2(1 + 2 + 3 + \cdots k)$$

and use the formula we derived on the previous page:

$$E = 2 \times \frac{k(k + 1)}{2} = k(k + 1)$$

Examples:

$$2 + 4 + 6 + \cdots + 88 = 44 \times 45$$

$$2 + 4 + 6 + \cdots + 100 = 50 \times 51$$

The sum of 2k − 1 consecutive odd integers starting at 1

Denote the sum with D:

$$D = 1 + 3 + 5 + \cdots + (2k - 1)$$

Add 1 to each term and notice that we obtain the sum E which we derived on the previous page:

```
D  =  1  +  3  +  5  +  ···  +  (2k-1)

      1  +  1  +  1  +  ···  +   1
   ─────────────────────────────────────  +
E  =  2  +  4  +  6  +  ···  +   2k
```

Since E consists of k multiples of 2, we must have added exactly k ones. Therefore,

$$
\begin{aligned}
D + k &= 2 + 4 + 6 + \cdots + (2k) \\
D + k &= k(k + 1) \\
D + k &= k^2 + k \\
D &= k^2
\end{aligned}
$$

We find that the sum of consecutive odd integers starting at 1 is always a perfect square. To find out which number is squared, add 1 to the last term and divide by 2.

Examples:

$$1 + 3 + 5 + \cdots + 21 = 11^2$$

$$1 + 3 + 5 + \cdots + 111 = 56^2$$

12

Calculating the sum of an arithmetic sequence can always be reduced to an application of the simplest arithmetic sequences we learned in the previous sections. One can always: subtract the same amount from each term, multiply/divide each term by the same number, etc. to reduce any arithmetic sum to a known sum.

Example: For an arithmetic sequence with first term 7 and common difference 3, what is the sum S:

$$S = 7 + 10 + 13 + \cdots + 37$$

Subtract 4 from each term and add it at the end:

$$S = 3 + 6 + 9 + \cdots + 33 + 4 + 4 + \cdots + 4$$

It is now obvious that there are 11 terms in each sequence. Take a factor of 3 in the first sequence and simplify the sequence of 4s:

$$S = 3(1 + 2 + \cdots + 11) + 4 \times 11$$

Apply the formula for the sum of the first 11 consecutive numbers:

$$
\begin{aligned}
S &= 3 \times \frac{11 \cdot 12}{2} + 44 \\
&= 3 \cdot 11 \cdot 6 + 44 \\
&= 18 \cdot 11 + 4 \cdot 11 = 22 \cdot 11 = 11^2 \cdot 2 = 242
\end{aligned}
$$

The sum and difference identity is often used in calculations, including in many problems based on sequences.

Use the distributive property to simplify:

$$(m + n)(m - n) = m^2 - nm + mn - n^2 = m^2 - n^2$$

Because the operations we have performed did not depend on any particular choices for the numbers m and n, we can say that, for any real numbers m and n it is true that:

$$m^2 - n^2 = (m + n)(m - n)$$

This is an identity, not an equation! In any computation we can replace $m^2 - n^2$ by $(m + n)(m - n)$ and conversely, depending on which form is more advantageous for the task at hand.

Example: In a right angle triangle, one leg has length 60 and the hypotenuse has length 61. What is the length of the other leg?

We have to apply the Pythagorean theorem. Denote the unknown length of the leg with x:

$$x^2 = 61^2 - 60^2$$

It is more difficult and time consuming to compute the two squares and subtract. However, the operation becomes easy if we apply the identity we just learned:

$$x^2 = (61 + 60)(61 - 60) = 121 = 11^2$$

PRACTICE ONE

Do not use a calculator for any of the problems!

Exercise 1

Calculate the sums:

(a) $1 + 2 + 3 + 4 + \cdots + 101 =$

(b) $1 + 2 + 3 + 4 + \cdots + 49 =$

(c) $1 + 2 + 3 + 4 + \cdots + 2p =$

(d) $1 + 2 + 3 + 4 + \cdots + (2m + 1) =$

(e) $m + (m + 1) + (m + 2) \cdots + (m + k) =$

Exercise 2

Calculate the sums:

(a) $2 + 4 + 6 + \cdots + 98 =$

(b) $2 + 4 + 6 + \cdots + 44 =$

(c) $2 + 4 + \cdots + 2p =$

(d) $22 + 44 + \cdots + 22p =$

(e) $2m + (2m + 2) + (2m + 4) \cdots + (2k) =$

Exercise 3

Calculate the sums:

(a) $1 + 3 + 5 + \cdots + 97 =$

(b) $1 + 3 + 5 + \cdots + 43 =$

(c) $1 + 3 + \cdots + (2p + 1) =$

(d) $23 + 25 + 27 + \cdots + 91 =$

(e) $(2m + 1) + (2m + 3) + (2m + 5) \cdots + (2k + 1) =$

Exercise 4

Calculate the expression:

$$X = 101 - 100 + 99 - 98 + \cdots + 1 =$$

Exercise 5

Calculate the expression:

$$P = 100 - 51 + 99 - 50 + 98 - 49 + \cdots + 50 - 1 =$$

Exercise 6

Calculate the expression:

$$G = 55 + 60 + 65 + 70 + 75 + \cdots + 235 =$$

Exercise 7

Calculate the expression:

$$Q = (2^2 - 1^2) + (3^2 - 2^2) + (4^2 - 3^2) + \cdots + (15^2 - 14^2) =$$

Exercise 8

Calculate the expression:

$$D = \frac{5^2}{1 + 3 + 5 + 7 + 9} =$$

Exercise 9

Calculate the expression:

$$T = \frac{19^2}{1 + 3 + 5 + \cdots + 37} =$$

Exercise 10

What must be the value of the positive integer number k that satisfies:

$$F = \frac{1 + 3 + 5 + \cdots + 59}{100k} = 1$$

Exercise 11

Find x. The overline is used to denote repeating decimals, i.e $2.\overline{4} = 2.4444\ldots.$

$$x = 2.\overline{3} + 3.\overline{6} + 4.\overline{3} + 5.\overline{6} + \cdots + 88.\overline{3} + 89.\overline{6}$$

Exercise 12

How many terms are the same in the following two sequences?

$$4, 9, 14, 19, 24, \ldots, 104$$

$$6, 13, 20, 27, \ldots, 167$$

Exercise 13

An arithmetic sequence has the common difference 4. What is the difference between the 5th and the 8th terms?

Exercise 14

A common geometric model of triangular numbers is the total numbers of small squares in the the 'triangular' parts of the construction:

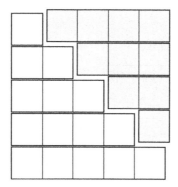

Notice how a perfect square is obtained by adding two consecutive triangular numbers. One of the triangular numbers differs from the other by the exact amount that is represented by the diagonal of the square.

Denote the k-th triangular number by T_k. Based on the geometric model, calculate:

$$T_{150} - T_{149}$$

as well as:

$$150^2 - 2 \cdot T_{149}$$

Exercise 15

A sequence of positive consecutive integers has a sum of 65. Which of the following cannot be the number of terms of the sequence?

(A) 2

(B) 5

(C) 10

(D) 12

Exercise 16

Calculate:

$$\frac{1 + 2 + 3 + \cdots + 100}{1 + 2 + 3 + \cdots + 99} =$$

Exercise 17

If T_k represents the k-th triangular number, which of the following identities is *not* correct?

(A) $T_{11} - T_{10} = 11$

(B) $12 + 15 + 18 + \cdots + 108 = 3T_{32} + 6T_{11}$

(C) $169 = T_{13} + T_{12}$

(D) $132 \cdot 23 = 11 \cdot T_{23}$

(E) $T_{19} + T_{18} = 18^2$

Exercise 18

Which of the following equalities is *not* true?

(A) $T_4 + T_5 = 1 + 3 + 5 + 7 + 9$

(B) $T_{11} = T_{12} - 12$

(C) $T_{11} + T_{12} = 12^2$

(D) $\dfrac{T_{12} + T_6}{2} = T_9$

Exercise 19

Match each identity to a geometric interpretation:

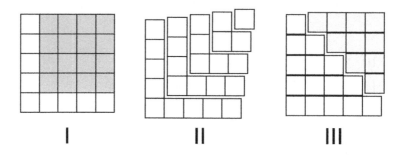

(A) $1 + 3 + 5 + \cdots + (2k - 1) = k^2$

(B) $T_{k-1} + T_k = k^2$

(C) $(k + 1)^2 - k^2 = 2k + 1$

Exercise 20

A sequence of stars is made using unit squares, as in the figure:

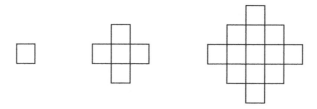

Find a simple way of calculating the number of unit squares in each star. Calculate the number of unit squares that must be used for the 25th star in the sequence.

Miscellaneous Sequences

In the **Fibonacci sequence** each term after the first two is the sum of its two predecessors:

$$0, 1, 1, 2, 3, 5, 8, 13, 21, 34, 55, 89, 144, \cdots$$

We will explore some of the properties of the Fibonacci sequence in the practice section.

The sequence formed by the **digit sums of positive integers** exhibits a pattern that is often used in problem statements:

$$1, 2, 3, \cdots, \mathbf{9}, \mathbf{1}, 2, \cdots, 9, \mathbf{10}, \mathbf{2}, 3, \cdots, 9, 10, \mathbf{11}, \mathbf{3}, 4, \cdots$$

The sequence formed by **the number of zeros at the end of a factorial** is another sequence with an interesting pattern:

$$0, 0, 0, 0, 0, 1, 1, 1, 1, 1, 2, 2, 2, 2, 2, 3, 3, 3, 3, 3, 4, 4, 4, 4, \mathbf{4}, \mathbf{6}, 6, 6, 6, 6, \cdots$$

Notice that no factorial ends with 5 zeros. Do you think there are other numbers of zeros that factorials cannot end in?

Other notable sequences are:

- The sequence of **perfect squares**, **perfect cubes**, and, in general, perfect powers.

- The sequence of **powers of 2**: $1, 2, 4, 8, 16, \cdots$.

- The sequence of **last digits** of the powers of an integer.

Experiment:

- Write the first 6 terms of the sequence of the last digit of each integer power of 7.

- Write the first 5 terms of the sequence of the last two digits of the integer powers of 5.

- Write the first 10 terms of the sequence of the last digit of factorials: $1!, 2!, 3!, \ldots$.

Do not use a calculator for any of the problems!

Exercise 1

Find the 46th term of the sequence of digit products of the non-negative consecutive integers starting at 0.

Exercise 2

What is the last digit of the number $N = 5^{1000} + 6^{1000}$?

Exercise 3

The rectangle in the figure has been completely dissected into squares. What percentage of the area of the rectangle has been shaded?

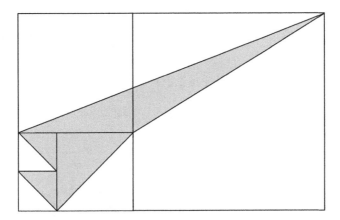

Exercise 4

Find the sum of the numbers whose factorials end in 11 zeros.

Exercise 5

The digit sums of two 3-digit numbers differ by 26. What is the digit sum of the sum of the numbers?

Exercise 6

The sequence of perfect squares starts with the following: $1^2, 2^2, \ldots$. Find the 700th term of the sequence of the last digits of perfect squares.

Exercise 7

The digit sum of the digit sum of the digit sum of the number N is 2. If we add 7 to the number we obtain:

(A) a multiple of 2

(B) a multiple of 3

(C) a multiple of 5

(D) a multiple of 6

(E) a multiple of 7

Exercise 8

Calculate the sum:

$$S = 1.\overline{1} + 2.\overline{2} + 3.\overline{3} + \cdots + 9.\overline{9}$$

Exercise 9

Find the last digit of

$$5^{2015} + 6^{2016} + 7^{2017} + 8^{2018} + 9^{2019}$$

24

Exercise 10

The sequence of perfect squares starts like this:

$$1, \ 2^2, \ 3^2, \ \cdots$$

How many of its first 20 terms are also perfect cubes?

Exercise 11

In base 2, numbers are written using only two digits: 1 and 0. Form a sequence of 8-digit numbers in base 2 with an odd number of odd digits in increasing order. Which is its 6th term?

Exercise 12

When simplified, the expression has the form of an irreducible fraction $\dfrac{n}{m}$:

$$\frac{1+2}{1+2+3} \times \frac{1+2+3+4}{1+2+3+4+5} \times \cdots \times \frac{1+2+\cdots+100}{1+2+3+\cdots+101}$$

Find $n + m$.

Exercise 13

Some integer numbers have the following digit sums:

(A) 78

(B) 1008

(C) 514

(D) 382

(E) 411

Which one of them could be followed by an integer with a digit sum of 1?

Exercise 14

Find the last digit of

$$2^{4k+3} + 3^{4k+1} + 4^{2k+1} + 5^k + 6^{k+1} + 7^{4k+2}$$

Exercise 15

The sequence of perfect squares is: $1^2, 2^2, 3^2, \ldots$. Is the number formed by writing the sequence of consecutive even integers from 2 to 64

$$2468101214 \cdots 606264$$

a term of this sequence?

Exercise 16

The position of a laser cutter changes by a unit step at each term of the sequence:

$$1, -1, -1, 1, -1, -1, 1, \cdots$$

The first step is horizontal and at each subsequent step the direction changes from horizontal to vertical, etc. Positive terms represent steps to the right or up, while negative terms represent steps to the left or down. What is the area that has been cut out after 745 steps?

Exercise 17

A sequence of grid-like assemblies formed of unit squares starts with the terms:

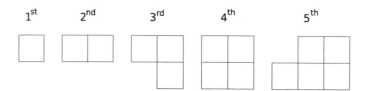

Unit squares are added in a clockwise spiral. What is the 25th term of the sequence formed by the corresponding perimeters? What is the 250th term of the sequence of perimeters?

Exercise 18

A sequence of segments of positive integer length has the property that no triangle of non-zero area can be built using any three of these segments. The largest segment length in this sequence is 144. What is the largest number of segments that can form such a sequence?

26

Exercise 19

All the positive integers with a digit product of 5 are written in increasing order from left to right. How many digits have been written as the 28th digit of 5 has been written?

Exercise 20

A number is formed of different digits arranged in decreasing order from left ro right. If a digit is on the right of an odd digit, then it can only be even. If a digit is on the right of an even digit, then it can be either even or odd. How many such numbers are there?

OPERATIONS WITH FRACTIONS

A **rational number** is a number of the form $\frac{a}{b}$, where a and b are integers and $b \neq 0$.

A set of equivalent fractions defines a **ratio**. Any rational number can be represented as a decimal number.

Any rational number q can be written as the sum of an integer and a proper fraction. The integer is called **the integer part** of q and the proper fraction is called **the fractional part** of q. The integer part of q is denoted by $[q]$. The fractional part of q is denoted by $\{q\}$ and we have: $\{q\} = q - [q]$.

It is always true that: $0 < \{q\} < 1$.

Example 1: If $q = 9.246$ then $[9.246] = 9$ and the fractional part is 0.246.

Example 2: If $q = -9.246$ then $[-9.246] = -10$ and the fractional part is 0.754.

Compare with the **ceiling** of a number:

Example 3: If $q = 9.246$ then $\lceil 9.246 \rceil = 10$.

Example 4: If $q = -9.246$ then $\lceil -9.246 \rceil = -9$.

and with the **floor** of a number:

Example 5: If $q = 9.246$ then $\lfloor 9.246 \rfloor = 9$.

Example 6: If $q = -9.246$ then $\lfloor -9.246 \rfloor = -10$.

To simplify **stacked fractions**, use the rules for operations with fractions and the order of operations. The order of operations can be inferred from the position of the '=' symbol. Be sure to align fraction lines correctly.

Example 1:

$$\frac{\frac{3}{4}}{\frac{9}{16}} = \frac{3}{4} \cdot \frac{16}{9} = \frac{4}{3}$$

Notice how examples 2,3 and 4 differ only by the length of the fraction line: the shorter the line, the higher the priority of the division. Translating to parentheses will clarify the notation:

Example 2:

$$\frac{\frac{3}{4}}{\frac{9}{16}} = \frac{\frac{3}{1} \cdot \frac{9}{4}}{16} = ((3 \div (4 \div 9)) \div 16)$$

Example 3:

$$\frac{\frac{3}{4}}{\frac{9}{16}} = \frac{\frac{3}{4} \cdot \frac{1}{9}}{16} = (((3 \div 4) \div 9) \div 16)$$

Example 4:

$$\frac{\frac{3}{4}}{\frac{9}{16}} = \frac{3}{\frac{4}{9} \cdot \frac{1}{16}} = (3 \div ((4 \div 9) \div 16))$$

Egyptian fractions, also called **unit fractions**, are fractions with numerator equal to 1.

Any fraction can be written as a sum of one or more Egyptian fractions.

Any unit fraction can be written as a sum of two different unit fractions, using the algebraic identity:

$$\frac{1}{n} = \frac{1}{n+1} + \frac{1}{n(n+1)}$$

However, this writing is not unique. For certain unit fractions, there are multiple ways in which they can be written as sums of two distinct unit fractions:

Examples: Write $\frac{1}{7}$ as a sum of two different unit fractions.

$$\frac{1}{7} = \frac{1}{21} + \frac{1}{42}$$
$$\frac{1}{7} = \frac{1}{8} + \frac{1}{56}$$

Practice Three

Do not use a calculator for any of the problems!

Exercise 1

Simplify:

$$\frac{13!}{12! \cdot 9} + \frac{25}{45} =$$

Exercise 2

Simplify:

$$(1000 \div 36) \div (125 \div 3) - (21 \div 56) \div (3 \div 4) + 25 \div 30 - 75 \times 4 \div 12 \div 100 =$$

Exercise 3

The overline is a notation for repeating non-terminating decimals. For example, $0.\overline{5} = 0.55555\ldots$.

Simplify:

$$\frac{1}{5}\left(19 - 3.\overline{3}\right) + \frac{1}{3}\left(\frac{3}{5} + \frac{7}{4 + \frac{1}{5}} + 3.\overline{3}\right)$$

Exercise 4

Calculate the value of the expression:

$$\frac{\dfrac{15}{4}}{9} \cdot \frac{\dfrac{8}{5}}{\dfrac{3}{2}} =$$

Exercise 5

What is the smallest number we have to multiply the following expression by in order to obtain a perfect square?

$$2 \times 3 \times 4 \times 5 \cdots \times 99 \div 2 \div 3 \div 4 \div \cdots \div 98$$

Exercise 6

Simplify:

$$\frac{11 \cdot 12 \cdot 13 \cdot 2002}{169 \cdot 242} =$$

Exercise 7

Simplify:

$$\left(\frac{1}{2} + \frac{1}{4} + \frac{1}{6}\right) \div \frac{121}{48} =$$

Exercise 8

In a set of 99 numbers, a number is divided by half of itself, another number is divided by a third of itself, another number is divided by a quarter of itself, and so on, until the last number is divided by one hundredth of itself. What is the sum of all the quotients?

Exercise 9

Simplify:

$$\frac{1}{6} + \frac{1}{12} + \frac{1}{20} + \cdots + \frac{1}{110} =$$

Exercise 10

Simplify:

$$\frac{2}{5 \cdot 7} + \frac{2}{7 \cdot 9} + \frac{2}{9 \cdot 11} + \cdots + \frac{2}{33 \cdot 35} =$$

Exercise 11

Simplify:

$$\frac{\frac{3}{2} + \frac{4}{3} + \frac{5}{4} + \cdots + \frac{44}{43} - \left(\frac{1}{2} + \frac{1}{3} + \cdots + \frac{1}{43}\right)}{42} =$$

Exercise 12

Simplify:

$$\frac{\frac{1}{6} + \frac{1}{10} + \frac{1}{15}}{\frac{1}{3} - \frac{1}{4}} \cdot \frac{\left(\frac{1}{2} + \frac{1}{26} + \frac{6}{13}\right) \cdot \frac{1}{4}}{\frac{1}{2} + \frac{1}{3} + \frac{1}{8} + \frac{1}{24}} =$$

Exercise 13

Simplify:

$$\frac{\left(0.2 + \frac{1}{70}\right) \cdot \frac{28}{9} \cdot \left(\frac{1}{2} + \frac{1}{6}\right)}{\left(\frac{1}{2} - \frac{1}{22}\right)\left(\frac{1}{3} + \frac{1}{30}\right)} =$$

Exercise 14

Use the identity:

$$\frac{1}{n} - \frac{1}{n+1} = \frac{1}{n(n+1)}$$

to write the fractions

  i. $\dfrac{3}{8}$

  ii. $\dfrac{4}{11}$

  ii. $\dfrac{2}{5}$

  iv. $\dfrac{7}{13}$

as sums of two Egyptian fractions with different denominators.

Exercise 15

How many different positive integers $1 \le m \le 19$ allow us to write the fraction $\dfrac{m}{20}$ as a sum of two non-trivial (i.e. not equal to 1) Egyptian fractions with different denominators?

Exercise 16

Calculate:

$$F = \frac{1}{2} + \frac{2}{3} + \frac{3}{4} + \cdots + \frac{2014}{2015} + 2 + \frac{3}{2} + \frac{4}{3} + \cdots + \frac{2016}{2015} =$$

Exercise 17

Find the value of the positive integer n if it is true that:

$$\frac{n}{n+1} = \frac{\left(\left(1 - \frac{1}{2} + \frac{1}{3}\right)\left(1 + \frac{1}{2} - \frac{1}{3}\right)\right) \div \left(\left(1 - \frac{1}{3} + \frac{1}{4}\right)\left(1 + \frac{1}{3} - \frac{1}{4}\right)\right)}{\left(\left(1 - \frac{1}{10} + \frac{1}{110}\right)\left(1 + \frac{1}{10} - \frac{1}{110}\right)\right)}$$

Exercise 18

Calculate:

$$\frac{5 + \frac{5}{6} - \left(\frac{1001}{13} - 66\right)\left(\frac{1}{2} - \frac{1}{3} - \frac{91}{1001}\right)}{6 + \frac{1}{4} - \frac{2}{5} + 3\left(\frac{1}{4} - \frac{6}{5}\right)} =$$

Exercise 19

Calculate:

$$\frac{18 \cdot 21 \cdot \left(307 + \frac{25}{2}\right)}{7 + 12 + 17 + \cdots + 632} =$$

Exercise 20

Calculate:

$$\frac{1}{709}\left(3 - \frac{1}{6} + 10 - \frac{2}{6} + 17 - \frac{3}{6} + 24 - \frac{4}{6} + \cdots + 360\right) =$$

Manipulating Integer Exponents in Problem Solving

Multiplying/dividing the same power of different bases can be written like this:

$$14^6 \times 13^6 = (14 \times 13)^6$$

$$14^6 \div 13^6 = \left(\frac{14}{13}\right)^6$$

Multiplying/dividing different powers of the same base can be written like this:

$$5^{13} \times 5^{11} = 5^{13+11} = 5^{24}$$

$$5^{13} \div 5^{11} = 5^{13-11} = 5^2$$

Adding different powers of different bases cannot be simplified:

$$2^5 + 5^2$$

If the exponent is the same, it is sometimes useful to factor out one of the powers in a sum:

$$5^7 + 3^7 = 3^7 \times \left(1 + \left(\frac{5}{3}\right)^7\right)$$

In a sum of different powers of the same base, it is sometimes useful to factor out the smallest power:

$$3^4 + 3^6 - 3^7 = 3^4 \cdot \left(1 + 3^2 - 3^3\right)$$

Since:

$$5^{11} \div 5^{13} = 5^{-2} = \frac{1}{5^2}$$

It means that a *negative exponent* is the same as applying the same power to the reciprocal number:

$$5^{-4} = \frac{1}{5^4}$$

$$\frac{1}{3^{-2}} = \frac{3^2}{1} = 3^2$$

It also follows that:

$$1 = 5 \div 5 = 5^1 \div 5^1 = 5^{1-1} = 5^0$$

Since this operation is valid for any non-zero base, we must have:

$$b^0 = 1$$

The base cannot be 0 **or** 1.

What to Do vs. What Not to Do:

Unlike exercises used in school to test the correct *application* of the order of operations (which is, supposedly, fixed), our exercises show the value of *creativity in planning an order of operations* (which is, actually, quite flexible.) They are excellent exercises in understanding the nature of the numbers, as well as of the operations.

As any creative activity, there is no single answer to how the operations should be executed. However, we do recommend that you consult the official solution at the back *for each and every problem.* Notice the differences between our solutions and yours (if any) and figure out the advantages/disadvantages of the one (or the other.)

These exercises are designed to increase your proficiency in working with numbers and will have far reaching effects on your overall math education. Please do not just look at them and decide they are boring drills, because they are not. Instead, see how they are similar to finding a good path when rock climbing. A calculator is like a bulldozer that will flatten the mountain for you, whereas you can be the nimble rock climber who gets to the ridge freely, and without damage.

- Do not apply PEMDAS in the order P, E, M, D, A, S from left to right. It will lead to complicated calculations with large numbers.

- Apply the order of operations *creatively* with the overall goal of keeping the numbers small.

- If possible, work with rational numbers, not with decimals.

- Start by factoring and dividing, not by multiplying.

- Avoid divisions by doing clever simplifications.

- Simplify expressions as much as possible before doing multiplications and additions. These are operations that make numbers large, therefore try to avoid them.

- Plan your next operation before doing it. Examine the expression and try various orders of operations before selecting an order that seems to reduce the size of the numbers, the number of terms, or the number of factors.

PRACTICE FOUR

Do not use a calculator for any of the problems!

Exercise 1
Simplify the operations. What is the exponent of 3?

$$3 \cdot 3^2 \cdot 3^3 \cdot 3^4 \cdot \ldots \cdot 3^{10}$$

Exercise 2
Simplify:

$$\frac{3^4 + 3^5}{4 \cdot 81}$$

Exercise 3
Simplify:

$$\frac{2^2 + 2^8}{4^2 + 4^5}$$

Exercise 4
Simplify:

$$\frac{25^4 + 5^6}{169} \cdot \frac{65}{10000}$$

Exercise 5
Simplify:

$$5^{2+4+6+\cdots+100} =$$

Exercise 6

What is the largest power of 3 that divides the expression:

$$T = 3^5 - 3^4 + 3^6 - 3^4 + 3^7 - 3^4$$

Exercise 7

Find the value of k:

$$\frac{2^{100}}{2} = (2^k)^{11}$$

Exercise 8

Which of the following equalities are not correct? Check all that apply.

(A)

$$-2^9 = (-2)^9$$

(B)

$$-2^8 = (-2)^8$$

(C)

$$2^{(-3)^3} = 2^{-3^3}$$

(D)

$$(-2)^{(-3)} = 2^3$$

Exercise 9

Find the smallest positive integers m and k such that:

$$\frac{6^k}{m} = 2^9 \cdot 3^6$$

Exercise 10

What is the difference between the smallest odd and the smallest even numbers that have a prime factorization of the form:

$$m^2 k^3$$

where m and k are different prime numbers?

Exercise 11

What is the largest power of 56 that divides:

$$(8^8 - 8^7)(8^7 - 8^6) \cdots (8^2 - 8)$$

Exercise 12

The result of the operations is a perfect cube of a number N. Find N.

$$N^3 = -3 \div 9 \times (-27) \div 81 \times (-243)$$

Exercise 13

Find the value of the positive integer k:

$$\frac{2^k + 6^k + 7^k + 10^k + 21^k + 35^k}{3^k + 5^k + 1} = 53$$

Exercise 14

Write the number E in base 3:

$$E = 3^8 - 3^7 + 3^6 - 3^5 + 3^4 - 3^3 + 3^2 - 3$$

Exercise 15

Find the positive integer k:

$$\frac{1}{k^5} = \frac{(3^{10} - 3^9)(3^8 - 3^7) \cdots (3^2 - 3)}{(3^2 - 2^3 + 1)^{10}(2^5 - 5)^{15}}$$

Exercise 16

Simplify:

$$\frac{21^2 \cdot 21^4 \cdots 21^{12}}{(2^3 - 1)^{21}(2^6 - 1)^{21}}$$

Exercise 17

How many prime divisors does the positive integer N have?

$$N = \sqrt{3^{1000} + 3^{1001} + 3^{1002} + 3^{1003} + 3^{1004}}$$

Exercise 18

Find the positive integer k such that:

$$3^{3^{3002}} \div 3^{3^{3001}} \div 3^{3^{3000}} = \left(3^{3^{3000}}\right)^k$$

Exercise 19

$\left(a^{b^2}\right)^{2b}$ is the square of which number?

Exercise 20

Which of the answer choices is equivalent to:

$$\frac{4!^{4!}}{5!^{5!}}$$

(A) $\dfrac{1}{5^{4! \cdot 4}}$

(B) $\dfrac{1}{5!^5}$

(C) $\dfrac{1}{5!^{5! \cdot 5}}$

(D) $\dfrac{1}{\left(5!^4 \cdot 5\right)^{4!}}$

(E) $\frac{1}{5^5}$

EXPRESSIONS

Operations with Factorials

Simplify factorials by using the factors they differ by:

$$\frac{8!}{9!} = \frac{8!}{8! \times 9} = \frac{\cancel{8!}}{\cancel{8!} \times 9} = \frac{1}{9}$$

Similarly, but with symbols:

$$\frac{N!}{(N-2)!} = \frac{\cancel{(N-2)!} \times (N-1) \times N}{\cancel{(N-2)!}} = (N-1) \times N$$

Similarly, when adding/subtracting:

$$k! - (k-1)! = (k-1)! \times (k-1)$$

Using Identities to Simplify Operations

The following algebraic identities are easy to prove and are useful in the process of performing complicated operations.

The Difference of Two Squares is an identity we have used in Chapter One, under the name of *sum and difference*. It is very useful when we have to factor numbers or expressions.

For any real numbers a and b it is true that:

$$a^2 - b^2 = (a + b)(a - b)$$

The proof is simple - we just have to use the distributive property of multiplication:

$$
\begin{aligned}
(a + b)(a - b) &= a(a - b) + b(a - b) \\
&= a^2 - ab + ba - b^2 \\
&= a^2 - b^2
\end{aligned}
$$

If a and b are integers, then $a + b$ and $a - b$ have the same *parity*, i.e. they are either both even or both odd. This is an useful observation if we are using this identity to solve a Diophantine equation.

If we want to calculate $199^2 - 190^2$ it is definitely easier to calculate:

$$
\begin{aligned}
199^2 - 190^2 &= (199 + 190)(199 - 190) = 389 \times 9 \\
&= 389 \times (10 - 1) = 3890 - 389 = 3501
\end{aligned}
$$

than to actually square the two numbers and subtract.

Perfect Squares, whether numeric or symbolic, can always be written as:

$$(m + n)^2 = m^2 + 2mn + n^2$$

or as:

$$(m - n)^2 = m^2 - 2mn + n^2$$

Any perfect square can be written in this form. Let us consider, for example, several different ways of writing 76^2:

$$
\begin{aligned}
76^2 &= (75 + 1)^2 = 75^2 + 2 \cdot 75 + 1 \\
76^2 &= (77 - 1)^2 = 77^2 - 2 \cdot 77 + 1 \\
76^2 &= (74 + 2)^2 = 74^2 + 2 \cdot 2 \cdot 74 + 2^2
\end{aligned}
$$

Think how difficult it would be to identify the perfect square if given in any of the forms on the right hand side.

Even more surprising is the application to irrational numbers:

$$(\sqrt{2} + \sqrt{3})^2 = 2 + 2 \cdot \sqrt{6} + 3 = 5 + 2\sqrt{6}$$

It is quite difficult to see that $5 + 2\sqrt{6}$ is the square of a sum of two terms.

The identity $(m + n)^2 = m^2 + 2mn + n^2$ can be visualized geometrically:

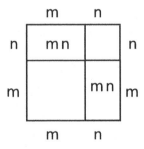

The large square with area $(m+n)^2$ can be dissected into two smaller squares with areas m^2 and n^2 and two rectangles with area $m \cdot n$.

In processing expressions, continue to follow the tenets:

1. **Attempt** to keep the numbers as small as possible, even if this results in an increased number of operations.

2. **Apply** the distributive property of addition/subtraction over multiplication correctly.

3. **Use** simple identities to simplify computations.

Notice that we may have to take decisions that exploit the flexibility of the order of operations to creating the most comfortable sequence of computations.

Example 1:

$$34 \times 98 = 34 \times (100 - 2) = 3400 - 68 = 3332$$

Example 2:

$$34 \times 55 + 34 \times 35 = 34 \times (55 + 35) = 34 \times 90 = 3060$$

Example 3:

$$105 - 92 + 55 - 38 = 105 + 55 - (92 + 38) = 160 - 130 = 30$$

When handling expressions, make sure you use the distributive property correctly when the factor is -1:

A factor of -1 changes *all* the signs in parenthesized expressions:

Example 1:

$$500 - (3 - 2 + 4 - 1) = 500 - 3 + 2 - 4 + 1$$

Example 2:

$$500 - (3 - (2 - 4 + 1)) = 500 - 3 + (2 - 4 + 1) = 500 - 3 + 2 - 4 + 1$$

Practice Five

Exercise 1

Find a rapid way to calculate:

$$M = 2016^2 - 2015 \cdot 2017$$

Exercise 2

Simplify:

$$\frac{5! + 6! + 7!}{49} =$$

Exercise 3

Find a rapid way to calculate:

$$2016^2 + 2015^2 - 2 \cdot 2015 \cdot 2016 =$$

Exercise 4

Calculate:

$$\Sigma = 2^2 - 1 \cdot 3 + 3^2 - 2 \cdot 4 + 4^2 - 3 \cdot 5 + \cdots + 2016^2 - 2015 \cdot 2017$$

Exercise 5

What is the digit sum of the number N:

$$N = 9 + 99 + 999 + 9999 + \cdots + \underbrace{999\ldots99}_{20 \text{ times}}$$

Exercise 6

Find the positive integer p if:

$$p^2 = \frac{1001! + 1002! + 1003!}{1003^2 \cdot 1001!}$$

Exercise 7

Find the smallest number we can multiply 3578575 by, in order to obtain a perfect square.

Exercise 8

Calculate the value of the expression:

$$10012^2 + 10011 - 10011^2 - 10012 =$$

Exercise 9

Simplify:

$$\frac{3^{10000} - 3^{9999}}{3^{9998}} \cdot \frac{1}{6} =$$

Exercise 10

Find the positive integer k such that:

$$\frac{10! + 12!}{19k} = 10!$$

Exercise 11

Calculate:

$$\sqrt{3 + \sqrt{3 + \sqrt{11}}} \cdot \sqrt{3 - \sqrt{3 + \sqrt{11}}} \cdot \sqrt{6 + \sqrt{11}} =$$

Exercise 12

Find the positive integer k such that:

$$\frac{k}{k+1} \cdot \frac{(k+1)!}{(k-1)!} = 17^2$$

Exercise 13

Simplify the left side of the equation:

$$(1 + 3 + 5 + \cdots + (2k - 1))(1 + 3 + 5 + \cdots + (2m - 1)) = 323^2$$

then find the positive numbers m and k.

Exercise 14

If x and y are positive integers with $y > 4$, is the fraction F irreducible?

$$\frac{(x + 1)(x + 4)}{(y - 1)(y - 4)}$$

Exercise 15

Simplify:

$$\sqrt{2016 \cdot 2018 + 1}$$

Exercise 16

If $a + b = 15$ and $ab = 3$, then what is the numerical value of:

$$E = \frac{a^2 - ab + b^2}{4a^2b + 4b^2a}$$

Exercise 17

Simplify the fraction:

$$F = \frac{4^{2016} - 4^{2015} + 4^{2014}}{9^{1006} + 9^{1007} + 9^{1008}}$$

Exercise 18

Simplify:

$$\frac{(j + 2)(j^2 + 4)(j^4 + 16)(j^8 + 256)}{(k + 2)(k^2 + 4)(k^4 + 16)(k^8 + 256)} \cdot \frac{k^{16} - 2^{16}}{j^{16} - 2^{16}}$$

Exercise 19

Compute the sum:

$$U = 1 \cdot 2 + 2 \cdot 3 + 3 \cdot 4 + \cdots + 10 \cdot 11$$

Exercise 20

Compute the sum:

$$T = 1 \cdot 2 \cdot 3 + 2 \cdot 3 \cdot 4 + 3 \cdot 4 \cdot 5 + \cdots + 10 \cdot 11 \cdot 12$$

INEQUALITIES

Very frequently used **numeric inequalities** are:

$$2^3 \; < \; 3^2 \qquad 8 < 9$$
$$2^2 \; > \; 3 \qquad 4 > 3$$
$$2^7 \; > \; 5^3 \qquad 128 > 125$$

Example with Numbers Larger than One

Which is larger 2^{49} or 5^{21}?

If $n \geq m > 1$ then, for any positive integer k, it is also true that $n^k \geq m^k$. Therefore,

$$2^7 \; > \; 5^3$$
$$(2^7)^7 \; > \; (5^3)^7$$
$$2^{49} \; > \; 5^{21}$$

Example with Numbers Smaller than One

Which is larger $\dfrac{1}{2^{49}}$ or $\dfrac{1}{5^{21}}$?

Since
$$2^{49} > 5^{21}$$

we have that:
$$\frac{1}{2^{49}} < \frac{1}{5^{21}}$$

The **arithmetic mean** of two numbers a and b is:

$$\text{AM} = \frac{a + b}{2}$$

The **geometric mean** of two numbers a and b is:

$$\text{GM} = \sqrt{a \cdot b}$$

The **arithmetic-mean geometric-mean (AM-GM) inequality** is the most popular inequality in problem statements. It is valid only for real numbers and is based on the fact that squares of real numbers are always non-negative. Then, it is also true that:

$$(\sqrt{x} - \sqrt{y})^2 \geq 0$$

Expanding, we obtain:

$$x - 2\sqrt{xy} + y \geq 0$$
$$\frac{x + y}{2} \geq \sqrt{xy}$$

which shows that the arithmetic mean of two real numbers is always greater than or equal to the geometric mean.

Both means can be defined for more than two numbers and the inequality can be proven to hold for any number of numbers.

For n numbers $\{x_1, x_2, \ldots, x_n\}$:

The **arithmetic mean** is:

$$AM = \frac{x_1 + x_2 + \cdots + x_n}{n}$$

The **geometric mean** is:

$$GM = \sqrt[n]{x_1 \cdot x_2 \cdots x_n}$$

The **arithmetic-mean geometric-mean (AM-GM) inequality** is

$$\frac{x_1 + x_2 + \cdots + x_n}{n} \geq \sqrt[n]{x_1 \cdot x_2 \cdots x_n}$$

The **triangle inequalities** are inequalities that a set of three numbers $\{a, b, c\}$ have to satisfy in order to be side lengths of a triangle:

$$
\begin{aligned}
a + b &> c \\
a + c &> b \\
b + c &> a
\end{aligned}
$$

PRACTICE SIX

Do not use a calculator for any of the problems!

Exercise 1
Which is greater 2^{27} or 3^{18}?

Exercise 2
Which is greater 2^{49} or 5^{35}?

Exercise 3
The numbers $A = 2^{63}$, $B = 5^{27}$, and $C = 3^{42}$ are ordered correctly as in:

(A) $B < A < C$

(B) $C < B < A$

(C) $B < C < A$

(D) $B < A < C$

Exercise 4
Identify the largest number in the sequence of fractions:

$$\frac{1}{52}, \frac{2}{54}, \frac{3}{56}, \ldots, \frac{99}{248}$$

Exercise 5
Which is larger: $\dfrac{3^{444}}{4}$ or $\dfrac{4^{355}}{3}$?

Exercise 6

Which number is larger A or B?

$$A = 1^1 + 2^2 + 3^2 + \cdots + 1000^2$$

$$B = 1 \cdot 2 + 2 \cdot 3 + 3 \cdot 4 + \cdots + 1000 \cdot 1001$$

Exercise 7

Calculate the value of the expression:

$$A = 25^{35} \div \left(\left| 2^{160} - 5^{70} \right| + 16^{40} \right)$$

Exercise 8

What is the cardinality (i.e. number of elements) of the set S:

$$S = \left\{ n \in \mathbb{Z} \middle| \frac{3 - \sqrt{7}}{2} < n < \frac{3 + \sqrt{7}}{2} \right\}$$

Exercise 9

Find the range of the natural numbers of the form $\sqrt{7ab6}$ where a and b are unknown digits of the 4-digit number $7ab6$.

Exercise 10

A triangle has integer side lengths of the form $x-3, x-1, 10-x$. How many possible values can x have?

Exercise 11

The sets A and B are defined as follows:

$$A = \left\{ a \in \mathbb{N} \middle| 2^{4671} \leq a < 2^{4672} \right\}$$

and

$$B = \left\{ b \in \mathbb{N} \middle| 5^{2001} \leq b < 5^{2002} \right\}$$

Which set has the larger cardinality (i.e. number of elements)?

Exercise 12

If a, b, and c are three different positive integers that satisfy $a < b < c$ and:

$$a(a - b)(a - c) = 23$$

find the difference $c - b$.

Exercise 13

If the digits a, b, c, d of a 4-digit number are distinct non-zero digits, what is the probability that this number does not have digits in increasing order or in decreasing order?

Exercise 14

How many pairs of positive integer numbers (a, b) are there such that $a+b < 6$ and $b - a < 3$?

Exercise 15

Given a 2-digit number ab such that $a > b$, its *reverse* is the number with the same digits but in reverse order: ba. What is the probability that, by subtracting the reverse from the number, we obtain a difference which is smaller than three times the sum of the two digits a and b?

Exercise 16

A number N is written in base 3 only with digits of 2. The largest number written in base 2 only with digits of 1 that is smaller than N has 11 digits. Find how many digits does the smallest such N have when written in base 3.

Exercise 17

True or false?

$$\frac{9}{10} \cdot \frac{11}{12} \cdot \frac{13}{14} \cdots \frac{99}{100} < \frac{3}{10}$$

Exercise 18

How many more digits than the number 4^{115} does the number 5^{230} have?

Miscellaneous Practice

Exercise 1

The expression is equivalent to the irreducible fraction $\dfrac{m}{n}$. Find $m + n$.

$$\frac{25^4 + 5^6}{169} \cdot \frac{65}{100000} =$$

Exercise 2

Simplify:

$$29^2 - 21^2 - 16^2 - 12^2 =$$

Exercise 3

Find the positive integer x:

$$65^2 - 60^2 - 24^2 - x^2 = 0$$

Exercise 4

Find the positive integer k:

$$\frac{(3^{10} - 3^9)(3^8 - 3^7) \cdot \cdots \cdot (3^2 - 3)}{(3^2 - 2^3 + 1)^5(2^5 - 5)^5} = 3^k$$

Exercise 5

Simplify the expression:

$$\frac{21^2 \cdot 21^4 \cdot \cdots \cdot 21^{12}}{(2^3 - 1)^{42}(2^3 + 1)^{21}} =$$

Exercise 6

A number is the geometric mean of the arithmetic means of 20 and 60 and of 13 and 32. Find the number.

Exercise 7

Calculate the arithmetic and geometric means of the numbers in the table. Which is larger?

n	m	$AM(m, n)$	$GM(m, n)$
10	40		
15	375		
12	48		
44	275		

Exercise 8

Simplify:
$$\frac{\text{lcm}(1, 2) \cdot \text{lcm}(3, 4) \cdots \text{lcm}(N, N + 1)}{(N + 1)!} =$$

Exercise 9

If a and b are positive integers, compute is the value of the expression:
$$(-1)^{ab^2 + a^2 b} + (-1)^{a(5a+1)} + (-1)^{(4a+b)(b+1)}$$

(A) 0

(B) 1

(C) 3

(D) depends on the values of a and b

Exercise 10

Find k:
$$1 + 3 + 5 + \cdots + k = 2^4 \cdot 3^2 \cdot 7^2$$

Exercise 11

Simplify:
$$G = \left(2 + \frac{3}{2} + \frac{4}{3} + \cdots + \frac{2016}{2015}\right) + \left(4030 - \frac{1}{2} - \frac{2}{3} - \cdots - \frac{2014}{2015}\right) =$$

Exercise 12

An arithmetic sequence starts with the terms: $n, k \cdot n$. For how many integer values of k is the fifth term of the sequence divisible by the third term?

Exercise 13

How many perfect squares smaller than 300 are perfect cubes?

Exercise 14

Damien the Magician has a hat full of cards. On each card there is a perfect square number. There are no two cards with the same number on them. Damien tells the audience he is able to take a few cards from the hat, blindfolded, so that at least two of the cards have numbers that differ by a multiple of 10. What is the smallest number of cards Damien must take out of the hat?

Exercise 15

Simplify:

$$\sqrt{4 + 2020 \sqrt{1 + 2017 \cdot 2015}}$$

Exercise 16

Find the sum of all the numbers k that can make the following expression an integer:

$$P = \sqrt{5 + \sqrt{3 + \sqrt{k}}} \cdot \sqrt{5 - \sqrt{3 + \sqrt{k}}} \cdot \sqrt{22 + \sqrt{k}}$$

Exercise 17

Simplify:

$$\frac{1 \cdot 2 \cdot 3 \cdot 4 \cdots \cdots 111}{1 \cdot 3 \cdot 5 \cdot 7 \cdot 9 \cdots \cdots 111} \cdot \frac{2^{-54}}{1 \cdot 2 \cdot 3 \cdots \cdots 55} =$$

Exercise 18

Find the positive integer n such that:

$$\frac{1}{1+2} + \frac{1}{1+2+3} + \frac{1}{1+2+3+4} + \cdots + \frac{1}{1+2+\cdots+n} = \frac{49}{50}$$

Exercise 19

If $x + \dfrac{1}{x} = 7$, find the positive value of $x - \dfrac{1}{x}$.

Solutions to Practice One

Do not use a calculator for any of the problems!

Solution 1

Notice how, throughout the computations, we choose operations that are simpler to perform, rather than a mechanical application of PEMDAS. The most important principle is to keep the numbers as small as possible. Other ideas may be used, of which a few can be noticed in the explicit computations rendered in the following:

(a) This is an arithmetic sequence with a first term of 1 and a common difference equal to 1. Apply the formula $N(N + 1)/2$ with $N = 101$:

$$
\begin{aligned}
1 + 2 + 3 + 4 + \cdots + 101 &= \frac{101 \times 102}{2} \\
&= 101 \times 51 \\
&= 5151
\end{aligned}
$$

(b) Apply the formula $N(N + 1)/2$ with $N = 49$:

$$
\begin{aligned}
1 + 2 + 3 + 4 + \cdots + 49 = \frac{49 \times \cancel{50}}{\cancel{2}} &= 49 \times 25 \\
&= (50 - 1) \times 25 \\
&= 1250 - 25 \\
&= 1225
\end{aligned}
$$

(c) Apply the formula $N(N + 1)/2$ with $N = 2p$:

$$
\begin{aligned}
1 + 2 + 3 + 4 + \cdots + 2p &= \frac{2p(2p + 1)}{2} \\
&= (2p + 1)p \\
&= 2p^2 + p
\end{aligned}
$$

(d) Apply the formula $N(N + 1)/2$ with $N = 2m + 1$:

$$
\begin{aligned}
1 + 2 + 3 + 4 + \cdots + (2m + 1) &= \frac{(2m + 1)(2m + 2)}{2} \\
&= \frac{2(2m + 1)(m + 1)}{2} \\
&= (2m + 1)(m + 1) \\
&= 2m^2 + 3m + 1
\end{aligned}
$$

(e) Rearrange the terms of the sum:

$$
m + m + m + \cdots + m + 1 + 2 + 3 + \cdots + k
$$

Notice that there is one more term equal to m as there are consecutive numbers added in the second part of the sum. That is, there are k terms equal to m whose sum is added to the sum of the consecutive numbers from 1 to k:

$$
m + m + m + \cdots + m + 1 + 2 + 3 + \cdots + k = (k + 1)m + \frac{k(k + 1)}{2}
$$

Solution 2

(a) This is an arithmetic sequence with a first term of 2 and a common difference equal to 2. Factor out a 2 to discover a more familiar sequence:

$$
\begin{aligned}
2 + 4 + 6 + \cdots + 98 &= 2(1 + 2 + 3 + \cdots + 49) \\
&= 2 \cdot \frac{49 \cdot 50}{2} \\
&= 49 \cdot 50 \\
&= 50 \cdot 50 - 50 \\
&= 2500 - 50 \\
&= 2450
\end{aligned}
$$

(b) This sum is similar to the previous one:

$$
\begin{aligned}
2 + 4 + 6 + \cdots + 44 &= 2(1 + 2 + 3 + \cdots + 22) \\
&= 2 \cdot \frac{22 \cdot 23}{2} \\
&= 22 \cdot 23 \\
&= 2 \cdot 11 \cdot 23 \\
&= 2 \cdot 253 \\
&= 506
\end{aligned}
$$

(c) This sum is similar to the previous one, except it is abstract:

$$
\begin{aligned}
2 + 4 + 6 + \cdots + 2p &= 2(1 + 2 + 3 + \cdots + p) \\
&= 2 \cdot \frac{p \cdot (p + 1)}{2} \\
&= p(p + 1) \\
&= p^2 + p
\end{aligned}
$$

(d) This sequence has a first term of 22 and a common difference of 22. Take a common factor of 22:

$$
\begin{aligned}
22 + 44 + \cdots + 22p &= 22(1 + 2 + 3 + \cdots + p) \\
&= 22 \cdot \frac{p \cdot (p + 1)}{2} \\
&= 11 \cdot p(p + 1)
\end{aligned}
$$

(e) This sequence has a first term of $2m$ and a common difference of 2. Rearrange the terms to form two different sums. The one sum is formed of terms that are all equal to $2m - 2$. The other sum is an arithmetic sequence starting at 2 and with a common difference of 2. The number of terms in both sequences is $k - m + 1$:

$$
\begin{aligned}
&(2m + 0) + (2m + 2) + \cdots + 2k \\
=\ & 2m - 2 + 2m - 2 + \cdots 2m - 2 + 2 + 4 + \cdots + 2(k - m + 1) \\
=\ & 2m - 2 + 2m - 2 + \cdots 2m - 2 + 2(1 + 2 + \cdots + (k - m + 1)) \\
=\ & (2m - 2) \cdot (k - m + 1) + 2\frac{(k - m + 1)(k - m + 2)}{2} \\
=\ & (k - m + 1)(2m - 2 + k - m + 2) \\
=\ & (k - m + 1)(k + m)
\end{aligned}
$$

Solution 3

(a) This is an arithmetic sequence with a first term of 1 and a common difference of 2. Add 1 to each term and take a common factor of 2. Do not forget to subtract the 1's:

$$
\begin{aligned}
1 + 3 + 5 + \cdots + 97 &= 2 + 4 + 6 + \cdots + 98 - (1 + 1 + \cdots + 1) \\
&= 2(1 + 2 + 3 + \cdots + 49) - 49 \\
&= 2 \cdot \frac{49 \cdot 50}{2} - 49 \\
&= 49 \cdot 50 - 49 \\
&= 49 \cdot (50 - 1) \\
&= 49 \cdot 49 \\
&= (50 - 1)^2 = 2500 - 100 + 1 \\
&= 2401
\end{aligned}
$$

You can also use the fact that the sum of k consecutive odd numbers starting at 1 is equal to k^2. You have to calculate the number of terms. The number of numbers is $\dfrac{97 + 1}{2} = 49$. The sum is equal to 49^2.

(b) This sum is similar to the previous one:

$$
\begin{aligned}
1 + 3 + 5 + \cdots + 43 &= 2 + 4 + 6 + \cdots + 44 - (1 + 1 + \cdots + 1) \\
&= 2(1 + 2 + 3 + \cdots + 22) - 22 \\
&= 2 \cdot \frac{22 \cdot 23}{2} - 22 \\
&= 22 \cdot 23 - 22 \\
&= 22 \cdot (23 - 1) \\
&= 22 \cdot 22 \\
&= 4 \cdot 121 \\
&= 484
\end{aligned}
$$

or use the fact that it is equal to the square of the number of terms. There are $\dfrac{43 + 1}{2} = 22$ terms in the sum. Therefore, the sum is equal to $22^2 = 121 \cdot 4 = 484$.

(c) This sum is similar to the previous one except it is abstract:

$$
\begin{aligned}
1 + 3 + 5 + \cdots + (2p + 1) &= 2 + 4 + 6 + \cdots + (2p + 2) - (1 + 1 + \cdots + 1) \\
&= 2(1 + 2 + 3 + \cdots + (p + 1)) - (p + 1) \\
&= 2 \cdot \frac{(p + 1) \cdot (p + 2)}{2} - (p + 1) \\
&= (p + 1) \cdot (p + 2) - (p + 1) \\
&= (p + 1) \cdot (p + 2 - 1) \\
&= (p + 1) \cdot (p + 1) \\
&= (p + 1)^2
\end{aligned}
$$

This exercise proves the formula.

(d) This sequence starts at 23 and has a common difference of 2. Split each term into a sum between 21 and another number. Then, group all the 21's together. There are as many 21's as terms in the remaining sequence of numbers.

$$
\begin{aligned}
23 + 25 + 27 + \cdots + 91 &= 21 + 2 + 21 + 4 + 21 + 6 + \cdots + 21 + 70 \\
&= 2 + 4 + 6 + \cdots + 70 + 21 + 21 + \cdots + 21 \\
&= 2 \cdot (1 + 2 + 3 + \cdots + 35) + 21 \cdot 35 \\
&= 35 \cdot 36 + 21 \cdot 35 \\
&= 35 \cdot (36 + 21) \\
&= 35 \cdot 57 \\
&= 1995
\end{aligned}
$$

(e) This arithmetic sequence starts at $2m + 1$ and has a common difference of 2. Separate $2m + 1$ from each term. There are as many $2m + 1$ as terms in the remaining sequence. Since in the remaining sequence, the first term is zero, there are $k - m + 1$ terms:

$$
\begin{aligned}
& (2m + 1) + (2m + 3) + (2m + 5) \cdots + (2k + 1) \\
= \ & 0 + 2 + 4 + \cdots + (2k + 1 - (2m + 1)) + (2m + 1) + \cdots + (2m + 1) \\
= \ & 2(1 + 2 + 3 + \cdots + (k - m)) + (2m + 1)(k - m + 1) \\
= \ & 2\frac{(k - m)(k - m + 1)}{2} + (2m + 1)(k - m + 1) \\
= \ & (k - m)(k - m + 1) + (2m + 1)(k - m + 1) \\
= \ & (k - m + 1)(k - m + 2m + 1) \\
= \ & (k - m + 1)(k + m + 1)
\end{aligned}
$$

Solution 4

Each pair of terms is equal to 1: $101 - 100, 99 - 98$, etc. Notice how all the odd terms are positive and the even terms are negative. This means that the last pair is $1 - 0$ but the zero term has been omitted. The positive terms effectively count the number of pairs. The positive terms are all odd numbers from 1 to 101 inclusive - a total of 51 numbers. Therefore, $X = 51$.

Solution 5

Each pair of terms has a sum of 49: $100-51, 99-50$, etc. The negative terms effectively count the number of pairs. There are 51 pairs. The expression is equal to:

$$P = 51 \times 49 = (50 + 1)(50 - 1) = 2500 - 1 = 2449$$

Notice how we have simplified the computations by using the identity:

$$(a + b)(a - b) = a^2 - b^2$$

Solution 6

This is an arithmetic sequence with a first term of 55 and a common difference of 5. Split 50 off of each term:

$$G = 50 + 5 + 50 + 10 + 50 + 15 + \cdots + 50 + 185$$

Regroup the terms:

$$G = 5 + 10 + 15 + \cdots + 185 + 50 + 50 + \cdots + 50$$

Factor out a 5 in the first group:

$$G = 5(1 + 2 + 3 + \cdots + 37) + 50 + 50 + \cdots + 50$$

The consecutive numbers effectively count the number of terms of the original sum. Therefore, there are 37 terms equal to 50. We have:

$$
\begin{aligned}
G &= 5 \cdot \frac{37 \cdot 38}{2} + 50 \cdot 37 \\
&= 5 \cdot 19 \cdot 37 + 50 \cdot 37 \\
&= 37(5 \cdot 19 + 50) \\
&= 37(95 + 50) \\
&= 37 \cdot 145 \\
&= 5365
\end{aligned}
$$

Solution 7

Notice that the terms of the sum *telescope* according to a pattern:

$$
\begin{aligned}
Q &= 2^2 - 1^2 + 3^2 - 2^2 + 4^2 - 3^2 + \cdots + 15^2 - 14^2 \\
&= 15^2 - 1 \\
&= 224
\end{aligned}
$$

Solution 8

Notice that the denominator is a sum of consecutive odd integers starting at 1. As such, it is equal to the square of the number of terms.

$$D = \frac{5^2}{1 + 3 + 5 + 7 + 9} = \frac{5^2}{5^2} = 1$$

Solution 9

Similarly to the previous exercise, the denominator is a sum of consecutive odd integers starting at 1. It is, however, a little more difficult to figure out how many terms are in the sum. To find the number of terms, add 1 to each term and factor out a 2: the consecutive numbers will effectively count the terms. There are $\dfrac{37+1}{2} = \dfrac{38}{2} = 19$ terms.

$$T = \frac{19^2}{1+3+5+\cdots+37} = \frac{19^2}{19^2} = 1$$

Solution 10

Multiply both sides of the equality by $100k$:

$$
\begin{aligned}
1+3+5+\cdots+59 &= 100k \\
\left(\frac{59+1}{2}\right)^2 &= 100k \\
30^2 &= 100k \\
9 &= k
\end{aligned}
$$

Solution 11

Use your knowledge of non-terminating repeating decimals. It is also useful to split the terms into an integer part and a fractional part:

$$x = 2+3+4+\cdots+89+\frac{3}{9}+\frac{3}{9}+\cdots+\frac{3}{9}+\frac{6}{9}+\frac{6}{9}+\cdots+\frac{6}{9}$$

The number of terms is given by the consecutive numbers, which range from 2 to 89. Therefore, there are 88 terms. Of these, half have a fractional

70

part of one third and half have a fractional part of two thirds:

$$
\begin{aligned}
x &= \frac{89 \cdot 90}{2} - 1 + 44 \cdot \frac{1}{3} + 44 \cdot \frac{2}{3} \\
&= 89 \cdot 45 - 1 + 44\left(\frac{1}{3} + \frac{2}{3}\right) \\
&= 89 \cdot 44 + 89 - 1 + 44 \\
&= 44 \cdot 90 + 88 \\
&= 44(90 + 2) \\
&= 44 \cdot 92 \\
&= 4048
\end{aligned}
$$

Solution 12

The first sequence is an arithmetic sequence with first term 4 and common difference 5. Let us call the sequence S and its general term of rank k is S_k. Any term S_k can be generated by substituting positive integer values for k in:

$$S_k = 4 + 5(k - 1)$$

The second sequence is an arithmetic sequence with first term 6 and common difference 7. Let us call the sequence T and its general term of rank p is T_p. Any term T_p can be generated by substituting positive integer values for p in:

$$T_p = 6 + 7(p - 1)$$

Terms from S can be equal to terms from T for any of the pairs of ranks k and p such that:

$$S_k = T_p$$

The question is now to find how many such pairs there are:

$$
\begin{aligned}
4 + 5(k - 1) &= 6 + 7(p - 1) \\
5k - 1 &= 7p - 1 \\
5k &= 7p
\end{aligned}
$$

This is a linear Diophantine equation (because k and p can have only integer values). The fundamental theorem of arithmetic says that if two numbers are the same, their prime factorizations must be identical. Because 5 and 7 are coprime (i.e. $\gcd(5, 7) = 1$), it follows from the theorem that k must be a multiple of 7 and p must be a multiple of 5.

We have to find out how many terms there are in the sequence S:

$$
\begin{aligned}
104 &= 4 + 5(k - 1) \\
20 &= k - 1 \\
k &= 21
\end{aligned}
$$

There are 21 terms in total.

Similarly, there are 24 terms in the sequence T:

$$
\begin{aligned}
167 &= 6 + 7(p - 1) \\
161 &= 7(p - 1) \\
p - 1 &= 161 \div 7 = 23
\end{aligned}
$$

There are:

$$
\left\lfloor \frac{21}{7} \right\rfloor = 3
$$

multiples of 7 within the range of k, and:

$$
\left\lfloor \frac{24}{5} \right\rfloor = 4
$$

multiples of 5 within the range of p, there are only 3 pairs (k, p) that generate identical terms in both sequences:

$$
(k, p) \in \{(7, 5), (14, 10), (21, 15)\}
$$

The corresponding terms are:

$$
\begin{aligned}
S_7 &= 4 + 5 \cdot 6 = 34 \\
T_5 &= 6 + 7 \cdot 4 = 34 \\
S_{14} &= 4 + 5 \cdot 13 = 69 \\
T_{10} &= 6 + 7 \cdot 9 = 69 \\
S_{21} &= 4 + 5 \cdot 20 = 104 \\
T_{14} &= 6 + 7 \cdot 14 = 104
\end{aligned}
$$

Solution 13

The difference between the 5th and the 8th terms is $4 \cdot 3 = 12$.

Solution 14

T_{150} and T_{149} can be modeled as the two parts obtained by dissecting a square of side 150 units. The difference between them is the number of unit squares on the diagonal:

$$T_{150} - T_{149} = 150$$

Also, while 150^2 is the area of a square with side 150 units, T_{149} is a triangular number based on this square, that does not include the diagonal:

$$150^2 - 2 \cdot T_{149} = 150$$

Solution 15

The answer is (D).

Since $65 = 5 \cdot 13$ is not a triangular number, the sequence must be the difference of two triangular numbers. Let us see if 65 can be the difference of two triangular numbers:

$$
\begin{aligned}
\frac{n(n + 1)}{2} - \frac{k(k + 1)}{2} &= 65 \\
n(n + 1) - k(k + 1) &= 130 \\
n^2 - k^2 + n - k &= 130 \\
(n + k)(n - k) + (n - k) &= 130 \\
(n - k)(n + k + 1) &= 2 \cdot 5 \cdot 13
\end{aligned}
$$

Since n and k are numbers of terms, they must have positive integer values. The equation is a Diophantine equation that can be solved by factoring. As it is already in factored form, we can list the solutions. Notice that $n + k + 1$ must be larger than $n - k$.

Case 1	$n + k + 1 = 130$ $n - k = 1$	$n = 65$ $k = 64$	1 term
Case 2	$n + k + 1 = 65$ $n - k = 2$	$n = 33$ $k = 31$	2 terms: 32 and 33.
Case 3	$n + k + 1 = 26$ $n - k = 5$	$n = 15$ $k = 10$	5 terms: 11, 12, 13, 14, 15.
Case 4	$n + k + 1 = 13$ $n - k = 10$	$n = 11$ $k = 1$	the sequence has 10 terms

Solution 16

This problem is simple practice of triangular numbers. Both the numerator and the denominator are triangular numbers:

$$\frac{1 + 2 + \cdots + 100}{1 + 2 + \cdots + 99} = \frac{\cancel{100} \cdot 101}{99 \cdot \cancel{100}} = \frac{101}{99}$$

Since 101 is prime, the fraction is irreducible.

Solution 17

Checking the validity of the identities is good practice of the properties of triangular numbers.

Since (A) and (C) are correct at first glance, let us examine (B). Separate a 12 out of each term and re-arrange terms:

$$12 + 15 + \cdots + 108 = 0 + 3 + 6 + \cdots + 96 + 12 + 12 + \cdots + 12$$

Take a factor of 3 in the first sum. Notice that the consecutive numbers effectively count the terms. There are 33 terms:

$$3(1 + 2 + \cdots + 32) + 12 \cdot 33 = 3 \cdot \frac{32 \cdot 33}{2} + 33 \cdot 12 = 3 \cdot (T_{32} + 11 \cdot 12)$$

This means that choice (B) is correct.

Let us process choice (D):

$$T_{23} = \frac{23 \cdot 24}{2} = 23 \times 12$$

$$11 \cdot T_{23} = 11 \cdot 12 \cdot 23 = 132 \cdot 23$$

which is correct. However, (E) is incorrect since $T_{19} + T_{18} = 19^2$.

Solution 18

Since the sum of consecutive odd numbers is the square of the number of numbers, choice (A) is correct:

$$T_4 + T_5 = 5^2 = 25$$

On the other hand:
$$1 + 3 + 5 + 7 + 9 = 5^2 = 25$$

Choice (B) is also correct, since it can be re-arranged as:

$$T_{12} - T_{11} = 12$$

and the difference of the consecutive triangular numbers is equal to the 'diagonal' of the square.

Choice (C) is obviously correct.

Choice (D) cannot be true because T_{12} is a multiple of 2 and T_6 is odd. Their sum is odd and their average results non-integer. T_9, however, as all triangular numbers has an integer value.

Solution 19

(A) corresponds with II.

(B) corresponds with III.

(C) corresponds with I.

Solution 20

Denote the area (number of unit squares) of the first star by S_1 and the area of the second star by S_2, etc. We have to calculate the area S_{25}.

One idea for simplifying the work is to dissect the star into triangular numbers. One way of doing this is illustrated in the figure below. Notice how, in the sequence of stars, the longest row of unit squares is always an odd number, consisting of the central square plus twice the number of squares on its left/right. Therefore, the 4th star will have 7 squares in its longest row of squares, etc.

The figure uses the 5th star in the sequence, which has 9 squares along its longest arms:

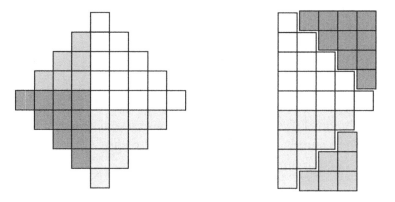

Notice how the triangular numbers can be recombined to form perfect squares. The area of the 5th star is:

$$S_5 = 5^2 + 4^2 = 25 + 16 = 31$$

Similarly, for the 25th star:

$$S_{25} = 25^2 + 24^2 = 625 + 576 = 1201$$

Solutions to Practice Two

Do not use a calculator for any of the problems!

Solution 1

The 46th term corresponds to the digit product of 45, which is 20.

Solution 2

Since all powers of numbers that end in 5 end in 5 and all powers of numbers that end in 6 end in 6, the last digit of the number N is the last digit of the sum $6 + 5$ - that is, 1.

Solution 3

Assume the side length of the smallest square is 1, then the lengths of the increasingly larger squares form a Fibonacci sequence:

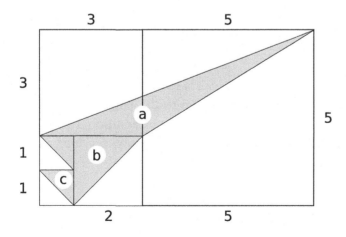

77

Therefore, the area of the triangle denoted by a is:

$$\frac{3 \times 3}{2} = \frac{9}{2} = 4.5$$

The area b is:

$$\frac{2 \times 2}{2} = 2$$

The area c is:

$$\frac{1 \times 1}{2} = 0.5$$

The total shaded area is:

$$A_{\text{shaded}} = 4.5 + 2 + 0.5 + 0.5 = 7.5$$

while the total area of the rectangle is:

$$A_{\text{total}} = 8 \times 5 = 40$$

and the desired percentage is:

$$\frac{A_{\text{shaded}}}{A_{\text{total}}} = \frac{7.5}{40} = \frac{75}{16 \times 25} = \frac{3}{16} = 18.75\%$$

Solution 4

A zero at the end of a number is produced by a pair of factors $(2, 5)$ in the prime factorization of the number. Therefore, the number of zeros at the end is equal to the largest number of pairs $(2, 5)$ that can be found in the prime factorization.

Because the number of factors of 2 is sufficiently large for any $k!$, the number of zeros at the end of $k!$ is equal to the number of factors of 5 in the prime factorization of $k!$ Since there is an additional factor of 5 every 5 numbers, an additional factor of 5 every 25 numbers, etc., the total number of factors of 5 is:

$$F = \left\lfloor \frac{k}{5} \right\rfloor + \left\lfloor \frac{k}{25} \right\rfloor + \cdots$$

where the sum continues until the fractions become smaller than 1.

Here is a table that summarizes the number of factors of 5 in the prime factorization of $k!$ as well as the number of zeros at the end of $k!$:

k	$k!$	number of factors of 5	number of zeros
1	1!	0	0
2	2!	0	0
3	3!	0	0
4	4!	0	0
5	5!	1	1
6	6!	1	1
...	...	...	...
10	10!	2	2
11	11!	2	2
...	...	...	...
15	15!	3	3
...	...	...	...
20	20!	4	4
...	...	...	...
25	25!	6	6
...	...	...	...
30	30!	7	7
...	...	...	...
35	35!	8	8
...	...	...	...
40	40!	9	9
...	...	...	...
45	45!	10	10
...	...	...	...
50	50!	12	12
...	...	...	...

There are no numbers whose factorial ends in 11 zeros! The answer is 0.

Solution 5

The largest digit sum that a 3-digit number can have is 27 and it is achieved if all the digits are equal to 9. If the digit sums of two 3-digit numbers differ by 26, one of them has to have a digit sum of 27 and the other has to have a digit sum of 1. Obviously, there are no 3-digit numbers with a digit sum of zero. Moreover, the only 3 digit number with a digit sum of 1 is 100. It turns out that, from the information given, we can infer both numbers: 999 and 100. The sum of these numbers is 1099 and has a digit sum of 19.

Solution 6

The last digits of perfect squares form the sequence:

$$1, 4, 9, 6, 5, 6, 9, 4, 1, 0, 1, 4, 9, 6, 5, \cdots$$

The terms of the sequence repeat every 10 numbers. The 700th term is equal to the 10th term: 0.

Solution 7

If the digit sum of the digit sum of a number is 2, then the number has a remainder of 2 when divided by 3 or by 9. If we add a 7 to the number, the sum is going to be a multiple of 9, as well as a multiple of 3. From the choices, (B) is true.

Solution 8

Use your knowledge of non-terminating repeating decimals:

$$S = 1 + \frac{1}{9} + 2 + \frac{2}{9} + \cdots + 9 + \frac{9}{9}$$

and re-organize the terms to separate them into two arithmetic sequences:

$$S = 1 + 2 + \cdots + 9 + \frac{1}{9}(1 + 2 + \cdots + 9)$$

Use the formula for triangular numbers and simplify the expression:

$$
\begin{aligned}
S &= (1 + 2 + \cdots + 9)\left(1 + \frac{1}{9}\right) \\
&= \frac{\cancel{9} \cdot 10}{2} \cdot \frac{10}{\cancel{9}} \\
&= 50
\end{aligned}
$$

Solution 9

All powers of 5 end with the digit 5 and all powers of 6 end with the digit 6.

The last digit of the powers of 7 forms the sequence:

$$7, 9, 3, 1, 7, 9, 3, 1, 7, \cdots$$

The terms of the sequence repeat every 4 terms. Since $2017 = 4 \times 504 + 1$, the 2017th term has the last digit 7.

The last digit of the powers of 8 forms the sequence:

$$8, 4, 2, 6, 8, 4, 2, 6, 8, \cdots$$

The terms of the sequence repeat every 4 terms. Since $2018 = 4 \times 504 + 2$, the 2018th term has the last digit 4.

The last digit of the powers of 9 forms the sequence:

$$9, 1, 9, 1, 9, \cdots$$

so that all the odd rank terms end in 9 and the even rank terms end in 1. The 2019th term has the last digit 9.

Since $5 + 6 + 7 + 4 + 9 = 31$, the last digit of the sum is 1.

Solution 10

For a perfect square to also be a perfect cube, it must be a perfect sixth power. The first 20 perfect squares range from 1 to 400. Since $512 = 2^9 = (2^3)^3 = 8^3$, the largest cube that is below 400 cannot be larger than 7^3. The only perfect square smaller than or equal to 7 is 4. Therefore 1 and:

$$2^6 = 4^3 = 64$$

are the only perfect sixth powers in the interval considered.

Solution 11

The smallest number has only one non-zero digit. The sequence continues directly with numbers that have three non-zero digits:

10000000

10000011

10000101

10000110

10001001

10001010

The 6th term is the binary 10001010.

Solution 12

$$\frac{1+2}{1+2+3} \times \frac{1+2+3+4}{1+2+3+4+5} \times \cdots \times \frac{1+2+\cdots+100}{1+2+3+\cdots+101} =$$

$$\frac{2 \cdot 3}{3 \cdot 4} \times \frac{4 \cdot 5}{5 \cdot 6} \times \cdots \times \frac{100 \cdot 101}{101 \cdot 102} =$$

$$\frac{2 \cdot \cancel{3}}{\cancel{3} \cdot \cancel{4}} \times \frac{\cancel{4} \cdot \cancel{5}}{\cancel{5} \cdot \cancel{6}} \times \cdots \times \frac{\cancel{100} \cdot \cancel{101}}{\cancel{101} \cdot 102} =$$

$$\frac{2}{102} = \frac{1}{51}$$

The answer is $1 + 51 = 52$.

Solution 13

Integers with a digit sum of 1 are powers of 10:

$$1, 10, 100, 1000, \ldots$$

Except for 1, the integers that precede them form the sequence of repdigits:

$$9, 99, 999, \ldots$$

The digit sums of the repdigits form the sequence of multiples of 9:

$$9, 18, 27, 36, 45, \ldots$$

The only answer choice that is a multiple of 9 is (B).

Solution 14

The last digit of powers of 2 forms the sequence:

$$2, 4, 8, 6, 2, 4, 8, 6, 2, \cdots$$

The terms of the sequence repeat every 4 terms. If the exponent is $4k + 3$, the respective power of 2 ends with the digit 8.

The last digit of powers of 3 forms the sequence:

$$3, 9, 7, 1, 3, 9, 7, 1, 3, \cdots$$

The terms of the sequence repeat every 4 terms. If the exponent is $4k + 1$, the respective power of 3 ends with the digit 3.

The last digit of powers of 4 forms the sequence:

$$4, 6, 4, 6, 4, \cdots$$

The terms of the sequence repeat every 2 terms. If the exponent is $2k + 1$, the respective power of 4 ends with the digit 4.

All powers of 5 end in 5 and all powers of 6 end with the digit 6.

The last digit of the powers of 7 forms the sequence:

$$7, 9, 3, 1, 7, 9, 3, 1, 7, \cdots$$

The terms of the sequence repeat every 4 terms. If the exponent is $4k + 2$, the respective power of 7 ends with the digit 9.

Since $8 + 3 + 4 + 5 + 6 + 9 = 35$, the last digit of the sum is 5 for any natural value of k.

Solution 15

Notice that the last three digits of the number form a number divisible by 8:

$$264 = 8 \times 33$$

Since $8 = 2^3$, for the number to be a perfect square it is necessary to be divisible by 16. Since:

$$6264 = 391 \times 16 + 8$$

the number is not divisible by 16 and is, therefore, not a perfect square.

Solution 16

Trace the first few steps:

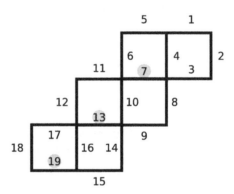

Notice that, after the first 4 steps, only 3 additional steps are needed to complete the next unit square. The pattern of unit squares continues down and to the left. The numbers of the sides marked with grey form an arithmetic sequence with common difference 6 and first term 7. Also, the number of complete squares cut after each step marked with grey forms the sequence: $2, 4, 6, \cdots$ which is an arithmetic sequence with common difference 2 and first term 2.

After 745 steps:

$$745 = 7 + 6 \times 123$$

there are 124 sides marked with grey. Therefore:

$$2 + 2 \times 123 = 248$$

there are 248 complete unit squares. The area that has been cut has 248 square units.

Solution 17

The sequence of perimeters starts with the terms:

$$4, 6, 8, 8, 10, 10, 12, 12, 12, 14, 14, 14, 16, 16, 16, 16, 18, 18, 18, 18, 20, 20, \ldots$$

The 25th term is equal to 20.

To calculate the 250th term, notice that the perimeter increases only *after* either a full square or a rectangle with sides $n-1$ and n has been completed. For example, let us look at the perimeter increases when going from the 3th term, which is a 1×2 rectangle, to the 4th term, or at the perimeter increase when going from the 4th term, which is a 2×2 square, to the 5th term. Also, the perimeter increases by 2 each time.

Notice that, for each perfect square of side n, the perimeter is equal to $4n$.

Since these are the only increases in perimeter, let us organize the counting on the base of the areas of complete squares and $(n-1) \times n$ rectangles. These areas form the sequence:

$$1, 2 = 1 \times 2, 4 = 2 \times 2, 6 = 2 \times 3, 9 = 3 \times 3, 12 = 3 \times 4, 16 = 4 \times 4, \ldots$$

and correspond to the following term ranks in the sequence of assemblies:

$$1^{\text{st}}, 2^{\text{nd}}, 4^{\text{th}}, 6^{\text{th}}, 9^{\text{th}}, 12^{\text{th}}, \ldots$$

Notice that, between these ranks the perimeters remain constant.

The perfect square that is closest to 250 is $256 = 16^2$. Therefore, the perimeter of the 256th assembly is $4 \times 16 = 64$ and it is also the perimeter of the 250th assembly.

Solution 18

If two segments have lengths that add up to the length of a third segment, the three segments cannot form a triangle. For example, 4, 5 and 9 cannot be the lengths of the sides of a triangle with non-zero area (non-degenerate triangle).

We can generate a sequence of side lengths that satisfy the requirement by ensuring that each succesive length is the sum of the previous two lengths:

$$1 + 1 = 2$$
$$1 + 2 = 3$$
$$2 + 3 = 5$$
$$\cdots$$

Notice that this is the Fibonacci sequence. Indeed, the number 144 is a term of the Fibonacci sequence starting with 1, 1:

$$1, 1, 2, 3, 5, 8, 13, 21, 34, 55, 89, 144, \ldots$$

This is also the largest possible number of such segment lengths. The answer is 12.

Solution 19

Positive integers that have a digit product of 5 have only digits equal to 1 and one digit equal to 5. In increasing order, they form the sequence:

$$5, 15, 51, 115, 151, 511, 1115, 1151, 1511, 5111, \ldots$$

The 28th digit of 5 occurs in the 28th term of the sequence. There are: 1 term that is a 1-digit number, 2 terms that are 2-digit numbers, and, in general, k terms that are k-digit numbers. Let us find out how many digits the 28th term has. Since:

$$1 + 2 + 3 + 4 + 5 + 6 + 7 = 28$$

the 28th term is the last term that is a 7-digit number: 5111111, and it starts with 5. Before this 5 has been written, a total of:

$$1 + 2 \cdot 2 + 3 \cdot 3 + 4 \cdot 4 + 5 \cdot 5 + 6 \cdot 6 = 91$$

digits have been written.

Solution 20

Let us build subsequences of numbers that satisfy the conditions, based on the leftmost digit. Each subsequence has the same leftmost digit and we increase the leftmost digit from 0 to 9 while we count.

Let us also denote each subsequence with a symbol such as $S_0, S_1, S_2, \ldots, S_9$, based on the value of the leftmost digit of the terms of the subsequence.

Let us also use the respective cardinalities $|S_0|, |S_1|, |S_2|, \ldots, |S_9|$, to denote the number of terms in each subsequence.

The subsequence S_0 starts with the digit 0. There is 1 such number. Therefore,

$$S_0 = 1$$

The subsequence S_1 starts with the digit 1, the number can only continue with a digit of 0. There are two such numbers that start with 1: 1 and 10. Notice that the number of terms in S_1 is equal to the sum of the numbers in the subsequence S_0 plus the single digit number 1:

$$|S_1| = 1 + |S_0| = 1 + 1 = 2$$

The subsequence S_2 consists of the number 2 (count 1 number), and continues with numbers that consist of appending all the previously derived numbers to the digit 2: $2, 20, 21, 210$. The number of terms in S_2 is equal to 1 (for the single digit number 2) plus the sum of the numbers of terms in all previous subsequences:

$$|S_2| = 1 + |S_0| + |S_1| = 1 + 1 + 2 = 4$$

The subsequence S_3 consists of the number 3 (count 1 number), and continues with numbers that consist of appending to 3 *only* the previously derived numbers that start with an even digit:

$$|S_3| = 1 + |S_0| + |S_2| = 1 + 1 + 4 = 6$$

The subsequence S_4 consists of the number 4 (count 1 number), and continues with numbers that consist of appending to 4 *all* the previously derived numbers:

$$|S_4| = 1 + |S_0| + |S_1| + |S_2| + |S_3| = 1 + 1 + 2 + 4 + 6 = 14$$

Similarly, for the subsequences S_5 through S_9 we have:

$$
\begin{aligned}
|S_5| &= |S_3| + |S_4| = 6 + 14 = 20 \\
|S_6| &= |S_4| + |S_4| + |S_5| = 2|S_4| + |S_5| = 2 \cdot 14 + 20 = 48 \\
|S_7| &= |S_5| + |S_6| = 20 + 48 = 68 \\
|S_8| &= 2|S_6| + |S_7| = 2 \cdot 48 + 68 = 96 + 68 = 164 \\
|S_9| &= |S_7| + |S_8| = 68 + 164 = 232
\end{aligned}
$$

The total number of required numbers is the sum of the cardinalities of all subsequences:

$$
1 + 2 + 4 + 6 + 14 + 20 + 48 + 68 + 164 + 232 = 559
$$

Solutions to Practice Three

Do not use a calculator for any of the problems!

Solution 1

Notice that $\dfrac{13!}{12!} = 13$:

$$\frac{13!}{12! \cdot 9} + \frac{25}{45} = \frac{13}{9} + \frac{5}{9}$$
$$= \frac{13 + 5}{9}$$
$$= 2$$

Solution 2

The expression can be re-arranged in fraction notation:

$$\frac{1000}{36} \times \frac{3}{125} - \frac{21}{56} \times \frac{4}{3} + \frac{25}{30} - \frac{75 \times 4}{12 \times 100}$$

$$= \frac{\cancel{125} \times 8}{\cancel{3} \times 12} \times \frac{\cancel{3}}{\cancel{125}} - \frac{\cancel{3} \times 7}{\cancel{4} \times 14} \times \frac{\cancel{4}}{\cancel{3}} + \frac{5}{6} - \frac{25 \times \cancel{12}}{\cancel{12} \times \cancel{25} \times 4}$$

$$= \frac{2}{3} - \frac{1}{2} + \frac{5}{6} - \frac{1}{4}$$

$$= \frac{8}{12} - \frac{6}{12} + \frac{10}{12} - \frac{3}{12}$$

$$= \frac{9}{12}$$

$$= \frac{3}{4}$$

89

Solution 3

$$\frac{1}{5}\left(19 - 3.\overline{3}\right) + \frac{1}{3}\left(\frac{3}{5} + \frac{7}{4 + \frac{1}{5}} + 3.\overline{3}\right) \; = \; \frac{1}{5}\left(19 - \frac{10}{3}\right) + \frac{1}{3}\left(\frac{3}{5} + \frac{7 \cdot 5}{21} + \frac{10}{3}\right)$$

$$= \; \frac{1}{5}\left(19 - \frac{10}{3}\right) + \frac{1}{3}\left(\frac{3}{5} + \frac{7 \cdot 5}{3 \cdot 7} + \frac{10}{3}\right)$$

$$= \; \frac{1}{5}\left(19 - \frac{10}{3}\right) + \frac{1}{3}\left(\frac{3}{5} + \frac{5}{3} + \frac{10}{3}\right)$$

$$= \; \frac{1}{5}\left(19 - \frac{10}{3}\right) + \frac{1}{3}\left(\frac{3}{5} + \frac{15}{3}\right)$$

$$= \; \frac{1}{5}\left(19 - \frac{10}{3}\right) + \frac{1}{3}\left(\frac{3}{5} + 5\right)$$

$$= \; \frac{1}{5}\left(\frac{57}{3} - \frac{10}{3}\right) + \frac{1}{3}\left(\frac{28}{5}\right)$$

$$= \; \frac{1}{5}\left(\frac{47}{3}\right) + \frac{28}{15}$$

$$= \; \frac{47}{15} + \frac{28}{15}$$

$$= \; \frac{47 + 28}{15}$$

$$= \; \frac{75}{15}$$

$$= \; 5$$

Solution 4

$$\frac{\dfrac{15}{4}}{9} \cdot \dfrac{8}{\dfrac{5}{\dfrac{3}{2}}} = \frac{15}{4} \cdot \frac{1}{9} \cdot \frac{8}{1} \div \dfrac{5}{\dfrac{3}{2}}$$

$$= \frac{15}{4} \cdot \frac{1}{9} \cdot \frac{8}{1} \div \left(\frac{5}{1} \cdot \frac{2}{3}\right)$$

$$= \frac{15}{4} \cdot \frac{1}{9} \cdot \frac{8}{1} \cdot \frac{1}{5} \cdot \frac{3}{2}$$

$$= \frac{\cancel{3} \cdot \cancel{5}}{\cancel{4}} \cdot \frac{1}{\cancel{3} \cdot \cancel{3}} \cdot \frac{\cancel{8}}{1} \cdot \frac{1}{\cancel{5}} \cdot \frac{\cancel{3}}{\cancel{2}} = 1$$

Solution 5

The expression can be simplified:

$$\frac{99!}{98!} = 99$$

The prime factorization of 99 is $3^2 \times 11$. The number must be multiplied by 11 to obtain the smallest perfect square that is a multiple of 99.

Solution 6

Notice that $2002 = 2 \times 1001$ and 1001 has the factorization:

$$1001 = 7 \cdot 11 \cdot 13$$

Also, notice that $169 = 13^2$ while 242 is a multiple of 11^2. Therefore:

$$\frac{11 \cdot 12 \cdot 13 \cdot 2002}{169 \cdot 242} = \frac{7 \cdot 11^2 \cdot 12 \cdot 13^2 \cdot 2}{13^2 \cdot 2 \cdot 11^2}$$

$$= 7 \cdot 12$$

$$= 84$$

91

Solution 7

Factor and divide rather than multiply.

$$
\left(\frac{1}{2}+\frac{1}{4}+\frac{1}{6}\right)\div\frac{121}{48} = \frac{6+3+2}{12}\cdot\frac{48}{11^2}
$$

$$
= \frac{\cancel{11}}{2\cdot\cancel{6}}\cdot\frac{\cancel{6}\cdot 8}{11\cancel{^2}}
$$

$$
= \frac{4}{11}
$$

Solution 8

We notice that, regardless of the number considered, dividing it by a half of itself yields the answer 2:

$$
\frac{N}{\frac{N}{2}} = \frac{N}{1}\cdot\frac{2}{N} = 2
$$

Likewise, a list of all the quotients is:

$$
2,3,4,\ldots,100
$$

and their sum is:

$$
\frac{100\cdot 101}{2} - 1 = 5050 - 1 = 5049
$$

Solution 9

Use Egyptian fractions:

$$\frac{1}{6} = \frac{1}{2} - \frac{1}{3}$$

$$\frac{1}{12} = \frac{1}{3} - \frac{1}{4}$$

$$\frac{1}{20} = \frac{1}{4} - \frac{1}{5}$$

$$\cdots$$

$$\frac{1}{110} = \frac{1}{10} - \frac{1}{11}$$

to turn the sum of fractions into a sum of differences that *telescope* (cancel out pairwise across consecutive terms):

$$\frac{1}{6} + \frac{1}{12} + \frac{1}{20} + \cdots + \frac{1}{110} =$$

$$\frac{1}{2} - \cancel{\frac{1}{3}} + \cancel{\frac{1}{3}} - \cancel{\frac{1}{4}} + \cancel{\frac{1}{4}} - \cancel{\frac{1}{5}} + \cdots + \cancel{\frac{1}{10}} - \frac{1}{11} =$$

$$\frac{1}{2} - \frac{1}{11} =$$

$$\frac{11 - 2}{22} =$$

$$\frac{9}{22}$$

Solution 10

Use a similar approach as in the previous exercise. Notice that the factors at the denominators are consecutive odd numbers, while the numerators are equal to 2. This observation gives us the idea:

$$\frac{2}{5 \cdot 7} = \frac{1}{5} - \frac{1}{7}$$

which applies similarly to all other terms of the sum. Since we managed to write each fraction as a difference we now hope that the sum will *telescope*:

$$\frac{2}{5 \cdot 7} + \frac{2}{7 \cdot 9} + \frac{2}{9 \cdot 11} + \cdots + \frac{2}{33 \cdot 35} =$$

$$\frac{1}{5} - \frac{1}{7} + \frac{1}{7} - \frac{1}{9} + \cdots + \frac{1}{33} - \frac{1}{35} =$$

$$\frac{1}{5} - \frac{1}{35} =$$

$$\frac{7 - 1}{35} =$$

$$\frac{6}{35}$$

Solution 11

Notice that:

$$\frac{3}{2} - \frac{1}{2} = 1$$

By re-arranging the terms at the numerator:

$$\frac{\frac{3}{2} - \frac{1}{2} + \frac{4}{3} - \frac{1}{3} + \cdots + \frac{44}{43} - \frac{1}{43}}{42}$$

$$= \frac{1 + 1 + 1 + \cdots + 1}{42}$$

$$= \frac{44 - 3 + 1}{42}$$

$$= 1$$

Solution 12

Work from left to right simplifying as soon as possible:

$$\frac{\frac{1}{6} + \frac{1}{10} + \frac{1}{15}}{\frac{1}{3} - \frac{1}{4}} \cdot \frac{\left(\frac{1}{2} + \frac{1}{26} + \frac{6}{13}\right) \cdot \frac{1}{4}}{\frac{1}{2} + \frac{1}{3} + \frac{1}{8} + \frac{1}{24}}$$

$$= \frac{\frac{5}{30} + \frac{3}{30} + \frac{2}{30}}{\frac{4}{12} - \frac{3}{12}} \cdot \frac{\left(\frac{13}{26} + \frac{1}{26} + \frac{12}{26}\right) \cdot \frac{1}{4}}{\frac{12}{24} + \frac{8}{24} + \frac{3}{24} + \frac{1}{24}}$$

$$= \frac{\frac{10}{30}}{\frac{1}{12}} \cdot \frac{\frac{26}{26} \cdot \frac{1}{4}}{\frac{24}{24}}$$

$$= \frac{1}{3} \cdot \frac{12}{1} \cdot \frac{1}{4}$$

$$= 1$$

Solution 13

Work from left to right, simplifying as much as possible:

$$\frac{\left(0.2 + \frac{1}{70}\right) \cdot \frac{28}{9} \cdot \left(\frac{1}{2} + \frac{1}{6}\right)}{\left(\frac{1}{2} - \frac{1}{22}\right)\left(\frac{1}{3} + \frac{1}{30}\right)}$$

$$= \frac{\left(\frac{1}{5} + \frac{1}{70}\right) \cdot \frac{28}{9} \cdot \left(\frac{3}{6} + \frac{1}{6}\right)}{\left(\frac{11}{22} - \frac{1}{22}\right)\left(\frac{10}{30} + \frac{1}{30}\right)}$$

$$= \frac{\left(\frac{14}{70} + \frac{1}{70}\right) \cdot \frac{28}{9} \cdot \left(\frac{3}{6} + \frac{1}{6}\right)}{\left(\frac{11}{22} - \frac{1}{22}\right)\left(\frac{10}{30} + \frac{1}{30}\right)}$$

$$= \frac{\frac{15}{70} \cdot \frac{28}{9} \cdot \frac{4}{6}}{\frac{10}{22} \cdot \frac{11}{30}}$$

$$= \frac{\frac{\cancel{5} \cdot \cancel{3}}{7 \cdot 2 \cdot \cancel{5}} \cdot \frac{7 \cdot 2 \cdot \cancel{2}}{\cancel{3} \cdot \cancel{3}} \cdot \frac{2}{3}}{\frac{\cancel{5}}{11} \cdot \frac{\cancel{11}}{\cancel{5} \cdot 6}}$$

$$= \frac{2 \cdot 2}{3 \cdot \cancel{3}} \cdot \frac{2 \cdot \cancel{3}}{1} = \frac{8}{3}$$

Solution 14

i.

$$\frac{3}{8} - \frac{3}{9} = \frac{3}{9 \cdot 8}$$

$$\frac{3}{8} - \frac{1}{3} = \frac{1}{24}$$

$$\mathbf{\frac{3}{8} = \frac{1}{3} + \frac{1}{24}}$$

ii.

$$\frac{4}{11} - \frac{4}{12} = \frac{4}{11 \cdot 12}$$

$$\frac{4}{11} - \frac{1}{3} = \frac{\cancel{4}}{11 \cdot 3 \cdot \cancel{4}}$$

$$\mathbf{\frac{4}{11} = \frac{1}{3} + \frac{1}{33}}$$

iii.

$$\frac{2}{5} - \frac{2}{6} = \frac{2}{5 \cdot 6}$$

$$\frac{2}{5} - \frac{1}{3} = \frac{\cancel{2}}{5 \cdot 3 \cdot \cancel{2}}$$

$$\mathbf{\frac{2}{5} = \frac{1}{3} + \frac{1}{15}}$$

iv.

$$\frac{7}{13} - \frac{7}{14} = \frac{7}{13 \cdot 14}$$

$$\frac{7}{13} - \frac{1}{2} = \frac{\cancel{7}}{13 \cdot 2 \cdot \cancel{7}}$$

$$\frac{7}{13} = \frac{1}{2} + \frac{1}{26}$$

Solution 15

This problem is a small research project since there is no theorem that allows us to predict if a fraction can be written or not as a sum of only two different unit fractions.

Since for $m \in \{1, 2, 4, 5, 10\}$ the fraction already has a numerator of 1, we can use the identity:

$$\frac{1}{n} = \frac{1}{n + 1} + \frac{1}{n(n + 1)}$$

to write them as sums of two unit fractions:

$$\frac{1}{2} = \frac{1}{3} + \frac{1}{6}$$

$$\frac{1}{4} = \frac{1}{5} + \frac{1}{20}$$

$$\frac{1}{5} = \frac{1}{6} + \frac{1}{30}$$

$$\frac{1}{10} = \frac{1}{11} + \frac{1}{110}$$

For $m = 3$ we have:

$$\frac{3}{20} - \frac{3}{21} = \frac{3}{20 \cdot 21}$$

$$\frac{3}{20} - \frac{1}{7} = \frac{\cancel{3}}{20 \cdot 7 \cdot \cancel{3}}$$

$$\frac{3}{20} = \frac{1}{7} + \frac{1}{140}$$

For $m = 6$ the fraction is reducible and we have:

$$\frac{3}{10} = \frac{2+1}{10}$$

$$\frac{3}{10} = \frac{2}{10} + \frac{1}{10}$$

$$\frac{3}{10} = \frac{1}{5} + \frac{1}{10}$$

For $m = 7$ we have:

$$\frac{7}{20} = \frac{5+2}{20}$$

$$\frac{7}{20} = \frac{5}{20} + \frac{2}{20}$$

$$\frac{7}{20} = \frac{1}{4} + \frac{1}{10}$$

For $m = 8$ the fraction is reducible and we have:

$$\frac{2}{5} - \frac{2}{6} = \frac{2}{5 \cdot 6}$$

$$\frac{2}{5} - \frac{1}{3} = \frac{\cancel{2}}{5 \cdot 3 \cdot \cancel{2}}$$

$$\frac{2}{5} = \frac{1}{3} + \frac{1}{15}$$

For $m = 9$ the fraction can be written as $\dfrac{9}{20} = \dfrac{4+5}{20} = \dfrac{1}{4} + \dfrac{1}{5}$.

For $m = 11$ we have:

$$\frac{11}{20} = \frac{1+10}{20}$$

$$\frac{11}{20} = \frac{1}{20} + \frac{1}{2}$$

For $m = 12$ the fraction is reducible:

$$\frac{3}{5} - \frac{3}{6} = \frac{3}{5 \cdot 6}$$

$$\frac{3}{5} - \frac{1}{2} = \frac{\cancel{3}}{5 \cdot 2 \cdot \cancel{3}}$$

$$\frac{3}{5} = \frac{1}{2} + \frac{1}{10}$$

For $m = 13$, at least three Egyptian fractions are required:

$$\frac{13}{20} = \frac{1}{3} + \frac{1}{4} + \frac{1}{15}$$

For $m = 14$:

$$\frac{7}{10} = \frac{2 + 5}{10}$$

$$\frac{7}{10} = \frac{2}{10} + \frac{5}{10}$$

$$\frac{7}{10} = \frac{1}{5} + \frac{1}{2}$$

For $m = 15$:

$$\frac{3}{4} = \frac{1 + 2}{4}$$

$$\frac{3}{4} = \frac{1}{4} + \frac{1}{2}$$

For $m \in 16, 17, 18, 19$ the fraction can be written using at least three Egyptian fractions.

There are 14 values of m that satisfy the conditions.

Solution 16

Notice that:

$$\frac{1}{2} + \frac{3}{2} = \frac{4}{2} = 2$$

and that:

$$\frac{2014}{2015} + \frac{2016}{2014} = \frac{2014 + 2016}{2015} = 2$$

The terms of the sum can be re-arranged in pairs that have a sum of 2. Find out how many pairs there are:

$$2015 - 2 + 1 = 2014$$

There are 2014 pairs to which we must also add the single term of 2. Therefore:

$$F = 2015 \cdot 2 = 4030$$

Solution 17

Simplify the right hand side:

$$
\begin{aligned}
G \;&=\; \frac{\left(\left(1-\frac{1}{2}+\frac{1}{3}\right)\left(1+\frac{1}{2}-\frac{1}{3}\right)\right) \div \left(\left(1-\frac{1}{3}+\frac{1}{4}\right)\left(1+\frac{1}{3}-\frac{1}{4}\right)\right)}{\left(1-\frac{1}{10}+\frac{1}{110}\right)\left(1+\frac{1}{10}-\frac{1}{110}\right)} \\[2mm]
&=\; \frac{\left(\left(1-\left(\frac{1}{2}-\frac{1}{3}\right)\right)\left(1+\left(\frac{1}{2}-\frac{1}{3}\right)\right)\right) \div \left(\left(1-\left(\frac{1}{3}-\frac{1}{4}\right)\right)\left(1+\left(\frac{1}{3}-\frac{1}{4}\right)\right)\right)}{\left(1-\left(\frac{1}{10}-\frac{1}{110}\right)\right)\left(1+\left(\frac{1}{10}-\frac{1}{110}\right)\right)} \\[2mm]
&=\; \frac{\left(\left(1-\frac{1}{6}\right)\left(1+\frac{1}{6}\right)\right) \div \left(\left(1-\frac{1}{12}\right)\left(1+\frac{1}{12}\right)\right)}{\left(1-\frac{1}{11}\right)\left(1+\frac{1}{11}\right)} \\[2mm]
&=\; \frac{\dfrac{5\cdot 7}{6^2} \div \dfrac{11\cdot 13}{12^2}}{\dfrac{10\cdot 12}{11^2}} \\[2mm]
&=\; \frac{5\cdot 7}{6^2}\cdot\frac{12^2}{11\cdot 13}\cdot\frac{11^2}{10\cdot 12} \\[2mm]
&=\; \frac{\cancel{5}\cdot 7}{6^2}\cdot\frac{12^{\cancel{2}}}{\cancel{11}\cdot 13}\cdot\frac{11^{\cancel{2}}}{2\cdot\cancel{5}\cdot\cancel{12}} \\[2mm]
&=\; \frac{7}{\cancel{12}\cdot 3}\cdot\frac{\cancel{12}}{13}\cdot\frac{11}{2} \\[2mm]
&=\; \frac{7}{3}\cdot\frac{1}{13}\cdot\frac{11}{2} \\[2mm]
&=\; \frac{77}{78}
\end{aligned}
$$

Therefore, $n = 77$

Solution 18

Work from left to right, identifying opportunities for simplification:

$$\frac{5 + \frac{5}{6} - \left(\frac{1001}{13} - 66\right)\left(\frac{1}{2} - \frac{1}{3} - \frac{91}{1001}\right)}{6 + \frac{1}{4} - \frac{2}{5} + 3\left(\frac{1}{4} - \frac{6}{5}\right)}$$

$$= \frac{5 + \frac{5}{6} - \left(\frac{7 \cdot 11 \cdot \cancel{13}}{\cancel{13}} - 66\right)\left(\frac{1}{2} - \frac{1}{3} - \frac{\cancel{7} \cdot \cancel{13}}{\cancel{7} \cdot 11 \cdot \cancel{13}}\right)}{6 + \frac{1}{4} + \frac{3}{4} - \frac{2}{5} - \frac{18}{5}}$$

$$= \frac{5 + \frac{5}{6} - (77 - 66)\left(\frac{1}{2} - \frac{1}{3} - \frac{1}{11}\right)}{6 + \frac{4}{4} - \frac{20}{5}}$$

$$= \frac{5 + \frac{5}{6} - 11\left(\frac{1}{6} - \frac{1}{11}\right)}{6 + 1 - 4}$$

$$= \frac{5 + \frac{5}{6} - 11\left(\frac{11}{66} - \frac{6}{66}\right)}{3}$$

$$= \frac{5 + \frac{5}{6} - \cancel{11} \cdot \frac{5}{6 \cdot \cancel{11}}}{3}$$

$$= \frac{5 + \frac{5}{6} - \frac{5}{6}}{3}$$

$$= \frac{5}{3}$$

Solution 19

Work from left to right, identifying opportunities for simplification. Notice the arithmetic sequence with common difference 5 and first term 7 at the denominator. Calculate its sum first:

$$
\begin{aligned}
7 + 12 + 17 + \cdots + 632 &= 7 + 7 + \cdots + 7 + 0 + 5 + 10 + 15 + \cdots 625 \\
&= 7 + 7 + \cdots + 7 + 5(1 + 2 + 3 + \cdots 125) \\
&= 7 \cdot 126 + 5 \cdot \frac{125 \cdot 126}{2} \\
&= 126\left(7 + \frac{625}{2}\right)
\end{aligned}
$$

The expression becomes:

$$
\begin{aligned}
\frac{18 \cdot 21 \cdot \left(307 + \frac{25}{2}\right)}{7 + 12 + 17 + \cdots + 632} &= \\
\frac{2 \cdot 3^3 \cdot 7\left(307 + \frac{25}{2}\right)}{126\left(7 + \frac{625}{2}\right)} &= \\
\frac{\cancel{2} \cdot 3^{\cancel{3}} \cdot \cancel{7}\left(307 + \frac{25}{2}\right)}{\cancel{2} \cdot 3^{\cancel{2}} \cdot \cancel{7}\left(7 + 300 + \frac{25}{2}\right)} &= 3
\end{aligned}
$$

Solution 20

Notice that the expression is the difference of two arithmetic sequences:

$$
\begin{aligned}
3 + 10 + 17 + 24 + \cdots + 360 &= 3 + 3 + 3 + \cdots + 3 + 0 + 7 + 14 + \cdots + 357 \\
&= 3 + 3 + 3 + \cdots + 3 + 0 + 7(1 + 2 + 3 + \cdots + 51) \\
&= 3 \cdot 52 + 7 \cdot \frac{51 \cdot 52}{2} \\
&= 52\left(3 + \frac{7 \cdot 51}{2}\right)
\end{aligned}
$$

We now know that the sequence of fractions has 51 terms since it has one less term than the sequence of integers:

$$
\begin{aligned}
\frac{1}{6} + \frac{2}{6} + \cdots + \frac{51}{6} &= \frac{1}{6}(1 + 2 + 3 + \cdots + 51) \\
&= \frac{1}{6} \cdot \frac{51 \cdot 52}{2}
\end{aligned}
$$

The difference is:

$$
\begin{aligned}
52\left(3 + \frac{7 \cdot 51}{2}\right) - \frac{1}{6} \cdot \frac{51 \cdot 52}{2} &= 52\left(3 + \frac{51}{2}\left(7 - \frac{1}{6}\right)\right) \\
&= 52\left(3 + \frac{\cancel{3} \cdot 17}{2} \cdot \frac{41}{2 \cdot \cancel{3}}\right) \\
&= 52\left(3 + \frac{17 \cdot 41}{4}\right) \\
&= \cancel{4} \cdot 13\left(\frac{12 + 17 \cdot 41}{\cancel{4}}\right) \\
&= 13(12 + 17 \cdot 41) \\
&= 13 \cdot 709
\end{aligned}
$$

And the expression is:

$$
\frac{1}{709} \cdot 13 \cdot 709 = 13
$$

SOLUTIONS TO PRACTICE FOUR

Do not use a calculator for any of the problems!

Solution 1

Notice that the sum of the exponents is a triangular number:

$$3 \cdot 3^2 \cdot 3^3 \cdot 3^4 \cdot \ldots \cdot 3^{10} = 3^{1+2+3+\cdots+10}$$

$$= 3^{\frac{10 \cdot 11}{2}}$$

$$= 3^{55}$$

Solution 2

Turn as many of the numbers as possible into powers of the same base. Then, use factoring to simplify:

$$\frac{3^4 + 3^5}{4 \cdot 81} = \frac{3^4 + 3^5}{4 \cdot 3^4}$$

$$= \frac{\cancel{3^4}(1 + 3)}{4 \cdot \cancel{3^4}}$$

$$= \frac{4}{4}$$

$$= 1$$

Solution 3

Turn as many of the numbers as possible into powers of the same base. Then, use factoring to simplify:

$$
\begin{aligned}
\frac{2^2 + 2^8}{4^2 + 4^5} &= \frac{2^2 + 2^8}{2^4 + 2^{10}} \\[2mm]
&= \frac{2^2(1 + 2^6)}{2^4(1 + 2^6)} \\[2mm]
&= \frac{2^6 + 1}{2^2(2^6 + 1)} \\[2mm]
&= \frac{1}{4}
\end{aligned}
$$

Solution 4

Simplify the expression:

$$
\frac{25^4 + 5^6}{169} \cdot \frac{65}{10000} = \frac{5^8 + 5^6}{169} \cdot \frac{65}{10000} =
$$

$$
\frac{5^6(25 + 1)}{13^2} \cdot \frac{5 \cdot 13}{2^4 \cdot 5^4} = \frac{5^6 \cdot 2 \cdot 13}{13^2} \cdot \frac{5 \cdot 13}{2^4 \cdot 5^4} =
$$

$$
\frac{5^3}{2^3} = \left(\frac{5}{2}\right)^3
$$

Solution 5

Notice that the exponent is an arithmetic sequence:

$$
\begin{aligned}
5^{2+4+6+\cdots+100} &= 5^{2(1+2+3+\cdots+50)} \\[2mm]
&= 5^{2 \cdot \frac{50 \cdot 51}{2}} = 5^{50 \cdot 51} \\[2mm]
&= 5^{2550}
\end{aligned}
$$

Solution 6

Notice that:

$$3^4 + 3^4 + 3^4 = 3 \cdot 3^4 = 3^5$$

Therefore, the expression can be simplified as:

$$T = 3^5 - 3^5 + 3^6 + 3^7 = 3^6 + 3^7$$

which has 6 factors of 3:

$$3^6 + 3^7 = 3^6(1 + 3) = 3^6 \cdot 4$$

Solution 7

Since:

$$\frac{2^{100}}{2} = 2^{100-1} = 2^{99}$$

we have:

$$2^{99} = 2^{11k}$$

and $k = 9$.

Solution 8

Verify each:

(A) is true

$$(-2)^9 = (-1)^9 \cdot 2^9 = -2^9$$

(B) is false

$$-2^8 \neq (-2)^8 = (-1)^8 \cdot 2^8 = 2^8$$

(C) is true

$$2^{(-3)^3} = 2^{(-1)^3 \cdot 3^3} = 2^{-3^3}$$

(D) is false

$$(-2)^{(-3)} = -\frac{1}{2^3} \neq 2^3$$

Solution 9

Since:

$$\frac{6^k}{m} = \frac{2^k \cdot 3^k}{m}$$

$$\frac{2^k \cdot 3^k}{m \cdot 2^9 \cdot 3^6} = 1$$

$$\frac{2^{k-9} \cdot 3^{k-6}}{m} = 1$$

$$2^{k-9} \cdot 3^{k-6} = m$$

The smallest value for k that makes m an integer is 9. Therefore $m = 3^3 = 27$.

Solution 10

The smallest odd number with the prime factorization $m^2 k^3$ is:

$$3^3 \cdot 5^2 = 27 \cdot 25 = \frac{2700}{4} = 675$$

The smallest even number with the prime factorization $m^2 k^3$ is:

$$2^3 \cdot 3^2 = 8 \cdot 9 = 72$$

The difference is $675 - 72 = 603$.

Solution 11

The answer is 7. Take factors of 8 in each of the parentheses:

$$(8^8 - 8^7)(8^7 - 8^6) \cdots (8^2 - 8) = 8^7(8-1) \cdot 8^6(8-1) \cdots 8(8-1)$$
$$= 8^{7+6+\cdots+1} \cdot 7^7$$

Since $56 = 8 \cdot 7$ and there are more than 7 factors of 8, the largest power of 56 that divides the expression is given by the largest power of 7 that divides the expression.

Solution 12

Put all the numbers into exponential form:

$$\begin{aligned} N^3 &= \frac{(-3) \times (-27) \times (-243)}{9 \times 81} \\ &= \frac{(-3)^{1+3+5}}{3^{2+4}} \\ &= \frac{(-3)^9}{3^6} \\ &= (-3)^3 \end{aligned}$$

Therefore, $N = -3$.

Solution 13

For any positive whole value of k, we can re-arrange the terms and attempt to factor:

$$\begin{aligned} \frac{2^k + 6^k + 10^k + 7^k + 21^k + 35^k}{1 + 3^k + 5^k} &= \frac{2^k\left(1 + 3^k + 5^k\right) + 7^k\left(1 + 3^k + 5^k\right)}{1 + 3^k + 5^k} \\ &= \frac{\left(1 + 3^k + 5^k\right)\left(2^k + 7^k\right)}{1 + 3^k + 5^k} \\ &= 2^k + 7^k \end{aligned}$$

For $k = 2$ the fraction has the value $2^2 + 7^2 = 4 + 49 = 53$.

Solution 14

Factor each pair of terms and use the fact that $2 = (3 - 1)$:

$$\begin{aligned} E &= 3^8 - 3^7 + 3^6 - 3^5 + 3^4 - 3^3 + 3^2 - 3 \\ &= 3^7(3 - 1) + 3^5(3 - 1) + 3^3(3 - 1) + 3(3 - 1) \\ &= 2 \cdot 3^7 + 2 \cdot 3^5 + 2 \cdot 3^3 + 2 \cdot 3 \\ &= 20202020_{(3)} \end{aligned}$$

Solution 15

Simplify the expression:

$$\frac{(3^{10} - 3^9)(3^8 - 3^7) \cdot \cdots \cdot (3^2 - 3)}{(3^2 - 2^3 + 1)^{10}(2^5 - 5)^{15}} = \frac{3^9 \cdot (3 - 1) \cdot 3^7 \cdot (3 - 1) \cdot \cdots \cdot 3(3 - 1)}{(9 - 8 + 1)^{10}(32 - 5)^{15}} =$$

$$\frac{3^9 \cdot 3^7 \cdot \cdots \cdot 3^1 \cdot 2^5}{2^{10}27^{15}} = \frac{3^{9+7+\cdots+1} \cdot 2^5}{2^{10}3^{45}} =$$

$$\frac{3^{25} \cdot 2^5}{2^{10}3^{45}} = \frac{1}{3^{20} \cdot 2^5} =$$

$$\frac{1}{(2 \cdot 3^4)^5} = \frac{1}{162^5}$$

Therefore, $k = 162$.

Solution 16

Simplify the expression:

$$\frac{21^2 \cdot 21^4 \cdot \cdots \cdot 21^{12}}{(2^3 - 1)^{21}(2^6 - 1)^{21}} = \frac{21^{2+4+\cdots+12}}{7^{21}((2^3 - 1)(2^3 + 1))^{21}} =$$

$$\frac{21^{2(1+2+\cdots+6)}}{7^{21}((8 - 1)(8 + 1))^{21}} = \frac{21^{2 \cdot \frac{6 \times 7}{2}}}{(7^2 \cdot 9)^{21}} =$$

$$\frac{21^{6 \times 7}}{7^{42} \cdot 9^{21}} = \frac{7^{42} \cdot 3^{42}}{7^{42} \cdot 3^{42}} = 1$$

Solution 17

Factor the sum:

$$
\begin{aligned}
N &= \sqrt{3^{1000}(1 + 3 + 3^2 + 3^3 + 3^4)} \\
&= 3^{500}\sqrt{1 + 3 + 9 + 27 + 81} \\
&= 3^{500}\sqrt{121} \\
&= 3^{500} \cdot 11
\end{aligned}
$$

N has two prime divisors: 3 and 11.

Solution 18

Calculate the left side:

$$
\begin{aligned}
\left(3^{3^{3002}} \div 3^{3^{3001}}\right) \div 3^{3^{3000}} &= 3^{3^{3002} - 3^{3001} - 3^{3000}} \\
&= 3^{3^{3000}(3^2 - 3 - 1)} \\
&= 3^{5 \cdot 3^{3000}} \\
&= \left(3^{3^{3000}}\right)^5
\end{aligned}
$$

and $k = 5$

Solution 19

Since:

$$
\begin{aligned}
a^{\left(b^2\right)^{2b}} &= a^{2b \cdot b^2} \\
&= a^{2 \cdot b^3} \\
&= \left(a^{b^3}\right)^2
\end{aligned}
$$

Solution 20

The correct answer is (D). Apply the properties of exponents:

$$\frac{4!^{4!}}{5!^{5!}} = \frac{4!^{4!}}{4!^{5!} \cdot 5^{5!}}$$

$$= \frac{4!^{4!}}{4!^{4! \cdot 5} \cdot 5^{5!}}$$

$$= \frac{1}{4!^{4!(5-1)} \cdot 5^{5!}}$$

$$= \frac{1}{4!^{4! \cdot 4} \cdot 5^{4! \cdot 5}}$$

$$= \frac{1}{(4!^4)^{4!} \cdot (5^5)^{4!}}$$

$$= \frac{1}{(4!^4 \cdot 5^5)^{4!}}$$

$$= \frac{1}{(5!^4 \cdot 5)^{4!}}$$

SOLUTIONS TO PRACTICE FIVE

Do not use a calculator for any of the problems!

Solution 1

Notice that $2015 = 2016 - 1$ and $2017 = 2016 + 1$:

$$
\begin{aligned}
2015 \cdot 2017 &= (2016 - 1)(2016 + 1) \\
&= 2016^2 - 1
\end{aligned}
$$

Therefore:

$$M = 2016^2 - (2016^2 - 1) = 1$$

Solution 2

$$
\begin{aligned}
\frac{5! + 6! + 7!}{49} &= \frac{5!(1 + 6 + 42)}{49} \\
&= \frac{5! \cdot \cancel{49}}{\cancel{49}} \\
&= 5! \\
&= 120
\end{aligned}
$$

Solution 3

Notice that the expression is a perfect square of the form:

$$a^2 + b^2 - 2ab = (a - b)^2$$

and use this identity to simplify:

$$2016^2 + 2015^2 - 2 \cdot 2015 \cdot 2016 = (2016 - 2015)^2 = 1$$

Solution 4

Notice that:

$$2 \cdot 4 = (3 - 1)(3 + 1) = 3^2 - 1$$

Similar numeric constructions are used throughout the expression.

$$2^2 - (2 - 1)(2 + 1) + \cdots + 2016^2 - (2016 - 1)(2016 + 1) =$$

$$2^2 - 2^2 + 1 + 3^2 - 3^2 + 1 + \cdots + 2016^2 - 2016^2 + 1 =$$

$$1 \cdot (2016 - 1) =$$

$$2015$$

The number of terms that are equal to 1 is equal to the number of differences. The differences range from 2 to 2016, therefore there are 2015 of them.

Solution 5

Use the fact that $9 - 1 = 10$ to rewrite the terms of the sum in a more convenient way:

$$N = 10 - 1 + 100 - 1 + 1000 - 1 + \cdots + 1\underbrace{000\ldots00}_{20 \text{ times}} - 1$$

Then, use the definition of place values to write the result in decimal form:

$$N = \underbrace{111\cdots11}_{20 \text{ times}}0 - 20$$

It is now easy to calculate N directly:

$$N = \underbrace{111\cdots11}_{18 \text{ times}}090$$

114

and the sum of the digits of N is:

$$S(N) = 18 + 9 = 27$$

Solution 6

Simplify the expression:

$$\frac{1001! + 1002! + 1003!}{1003^2 \cdot 1001!} = \frac{\cancel{1001!}(1 + 1002 + 1002 \cdot 1003)}{1003^2 \cancel{1001!}}$$

$$= \frac{1003 + 1002 \cdot 1003}{1003^2}$$

$$= \frac{1003(1 + 1002)}{1003^2}$$

$$= \frac{1003^2}{1003^2}$$

$$= 1$$

to find that $p = 1$.

Solution 7

In the factorization of a perfect square, all the factors are even powers. We begin by factoring 3578575. We notice that the digits 3, 5, and 7 occur at a distance of 4 place values and that $8 = 5 + 3$. This gives us the idea of writing the number like this:

$$3003000 + 500500 + 70070 + 5005$$

a sum of which all terms are multiples of 1001. Factor out a $1001 = 7 \cdot 11 \cdot 13$:

$$7 \cdot 11 \cdot 13(3000 + 500 + 70 + 5) = 7 \cdot 11 \cdot 13 \cdot 3575$$

Since 3575 is divisible by 25, we can rapidly factor it into:

$$3575 = 5^2 \times 11 \times 13$$

Therefore, the number is:

$$3578575 = 5^2 \times 7 \times 11^2 \times 13^2$$

It is sufficient to multiply it by 7 to obtain a perfect square.

Solution 8

Re-arrange the terms more conveniently:

$$10012^2 + 10011 - 10011^2 - 10012 = 10012^2 - 10012 - (10011^2 - 10011)$$

Use factoring:

$$10012^2 - 10012 - (10011^2 - 10011)$$

$$= 10012(10012 - 1) - 10011(10011 - 1)$$

$$= 10012 \cdot 10011 - 10011 \cdot 10010$$

$$= 10011(10012 - 10010)$$

$$= 10011 \cdot 2$$

$$= 20022$$

Or use the difference of squares for an alternate solution.

Solution 9

Use factoring to reduce the size of the numbers:

$$\frac{3^{10000} - 3^{9999}}{3^{9998}} \cdot \frac{1}{6} = \frac{3^{9999}(3 - 1)}{3^{9998}} \cdot \frac{1}{2 \cdot 3}$$

$$= \frac{3 \cdot 2}{2 \cdot 3}$$

$$= 1$$

Solution 10

Simplify:

$$\frac{10! + 12!}{19k} = \frac{10!(1 + 11 \cdot 12)}{19k}$$

$$= \frac{10! \cdot 133}{19k}$$

$$= \frac{10! \cdot 7 \cdot \cancel{19}}{\cancel{19}k}$$

$$= \frac{10! \cdot 7}{k}$$

If $k = 7$ the expression is equal to $10!$

Solution 11

Use the identity $(a + b)(a - b) = a^2 - b^2$:

$$\sqrt{\left(3 + \sqrt{3 + \sqrt{11}}\right) \cdot \left(3 - \sqrt{3 + \sqrt{11}}\right)} \cdot \sqrt{6 + \sqrt{11}}$$

$$= \sqrt{9 - (3 + \sqrt{11})} \cdot \sqrt{6 + \sqrt{11}}$$

$$= \sqrt{6 - \sqrt{11}} \cdot \sqrt{6 + \sqrt{11}}$$

$$= \sqrt{(6 - \sqrt{11})(6 + \sqrt{11})}$$

$$= \sqrt{36 - 11}$$

$$= \sqrt{25}$$

$$= 5$$

Solution 12

Simplify the factorials:

$$\frac{k}{k+1} \cdot \frac{(k+1)!}{(k-1)!} = \frac{k}{k+1} \cdot \frac{\cancel{(k-1)!} \cdot k \cdot (k+1)}{\cancel{(k-1)!}}$$

$$= \frac{k}{\cancel{(k+1)}} \cdot k \cdot \cancel{(k+1)}$$

$$= k^2$$

We find that $k = 17$.

Solution 13

Both parentheses contain sums of consecutive odd integers. There are k terms in the first sum and m terms in the second sum. Therefore, the left hand side is equal to $k^2 \cdot m^2$. The equation becomes:

$$k^2 \cdot m^2 = 323^2$$

and

$$k \cdot m = 323$$

Since k and m are numbers of terms, they are positive integers. This means we have a Diophantine equation to solve. Let us try factoring as a first attempt to solve it:

$$k \cdot m = 17 \cdot 19$$

There are two solutions:

$$k = 17, \quad \text{and} \quad m = 19$$

or

$$k = 1, \quad \text{and} \quad m = 323$$

Solution 14

If x is even, then $x+1$ is odd and $x+4$ is even. If x is odd, then $x+1$ is even and $x+4$ is odd. In both cases, $(x+1)(x+4)$ is an even number.

By a similar argument, $(y-1)(y-4)$ is also even for any value of y.

Therefore, the fraction is reducible by 2 for any values of x and y allowed by the statement.

Solution 15

Use the observation that 2016 and 2018 differ by 2:

$$
\begin{aligned}
\sqrt{2016 \cdot 2018 + 1} &= \sqrt{(2017-1)(2017+1)+1} \\
&= \sqrt{2017^2 - 1 + 1} \\
&= \sqrt{2017^2} \\
&= 2017
\end{aligned}
$$

Solution 16

Write the expression in terms of $a+b$ and ab:

$$
\begin{aligned}
E &= \frac{a^2 - ab + b^2}{4a^2 b + 4b^2 a} \\
&= \frac{a^2 + b^2 + 2ab - 3ab}{4ab(a+b)} \\
&= \frac{(a+b)^2 - 3ab}{4ab(a+b)}
\end{aligned}
$$

and substitute the numerical values given:

$$
\begin{aligned}
E &= \frac{15^2 - 3 \cdot 3}{4 \cdot 3 \cdot 15} \\
&= \frac{15^2 - 3^2}{2^2 \cdot 3^2 \cdot 5} \\
&= \frac{(15 - 3)(15 + 3)}{2^2 \cdot 3^2 \cdot 5} \\
&= \frac{12 \cdot 18}{2^2 \cdot 3^2 \cdot 5} \\
&= \frac{2^3 \cdot 3^3}{2^2 \cdot 3^2 \cdot 5} \\
&= \frac{6}{5} \\
&= 1.2
\end{aligned}
$$

Solution 17

Factor:

$$
\begin{aligned}
F &= \frac{4^{2014}\left(4^2 - 4 + 1\right)}{9^{1006}\left(1 + 9 + 9^2\right)} \\
&= \frac{4^{2014} \cdot 13}{3^{2012} \cdot 91} \\
&= \frac{4^{2014} \cdot \cancel{13}}{3^{2012} \cdot \cancel{13} \cdot 7} \\
&= \frac{4^{2014}}{7 \cdot 3^{2012}}
\end{aligned}
$$

Solution 18

Multiply by $\dfrac{j-2}{k-2}$ and apply the identity $(m+n)(m-n) = m^2 - n^2$:

$$\frac{(j-2)}{(k-2)} \frac{(j+2)(j^2+4)(j^4+16)(j^8+256)}{(k+2)(k^2+4)(k^4+16)(k^8+256)} \cdot \frac{k^{16}-2^{16}}{j^{16}-2^{16}} =$$

$$\frac{(j^2-4)(j^2+4)(j^4+16)(j^8+256)}{(k^2-4)(k^2+4)(k^4+16)(k^8+256)} \cdot \frac{k^{16}-2^{16}}{j^{16}-2^{16}} =$$

$$\frac{(j^4-16)(j^4+16)(j^8+256)}{(k^4-16)(k^4+16)(k^8+256)} \cdot \frac{k^{16}-2^{16}}{j^{16}-2^{16}} =$$

$$\frac{(j^8-256)(j^8+256)}{(k^8-256)(k^8+256)} \cdot \frac{k^{16}-2^{16}}{j^{16}-2^{16}} =$$

$$\frac{j^{16}-2^{16}}{k^{16}-2^{16}} \cdot \frac{k^{16}-2^{16}}{j^{16}-2^{16}} = 1$$

The original expression is equal to:

$$\frac{(j+2)(j^2+4)(j^4+16)(j^8+256)}{(k+2)(k^2+4)(k^4+16)(k^8+256)} \cdot \frac{k^{16}-2^{16}}{j^{16}-2^{16}} = \frac{k-2}{j-2}$$

Solution 19

It is possible to write a product of 2 consecutive numbers as a difference of products of 3 consecutive numbers:

$$k(k+1)(k+2) - (k-1)k(k+1) \;=\;$$

$$k(k+1)(k+2 - (k-1)) \;=\;$$

$$k(k+1)(3)$$

This observation leads us to an identity of the form:

$$k(k+1) = \frac{1}{3}\,(k(k+1)(k+2) - (k-1)k(k+1))$$

The sum U can be telescoped:

$$U = \frac{1}{3}(1\cdot2\cdot3 - 0\cdot1\cdot2 + 2\cdot3\cdot4 - 1\cdot2\cdot3 + 3\cdot4\cdot5 - 2\cdot3\cdot4 + \cdots + 10\cdot11\cdot12 - 9\cdot10\cdot11)$$

and calculated as:

$$U = \frac{1}{3}(10 \cdot 11 \cdot 12 - 0)$$

$$U = \frac{1}{3}(1320)$$

$$U = 440$$

Solution 20

It is possible to write a product of 3 consecutive numbers as a difference of products of 4 consecutive numbers:

$$k(k + 1)(k + 2)(k + 3) - (k - 1)k(k + 1)(k + 2) =$$

$$k(k + 1)(k + 2)(k + 3 - (k - 1)) =$$

$$k(k + 1)(k + 2)(4)$$

This observation leads us to an identity of the form:

$$k(k + 1)(k + 2) = \frac{1}{4}(k(k + 1)(k + 2)(k + 3) - (k - 1)k(k + 1)(k + 2))$$

The sum T can be telescoped:

$$T = \frac{1}{4}(1 \cdot 2 \cdot 3 \cdot 4 - 0 + 2 \cdot 3 \cdot 4 \cdot 5 - \cdots + 10 \cdot 11 \cdot 12 \cdot 13 - 9 \cdot 10 \cdot 11 \cdot 12)$$

$$T = \frac{1}{4}(10 \cdot 11 \cdot 12 \cdot 13)$$

$$T = 13 \cdot 33$$

$$T = 4290$$

Solutions to Practice Six

Do not use a calculator for any of the problems!

Solution 1

Since $2^3 < 3^2$ $(8 < 9)$:

$$\left(2^3\right)^9 \; < \; \left(3^2\right)^9$$

$$2^{27} \; < \; 3^{18}$$

Solution 2

Since $125 < 128$, we can use the fact that $2^7 > 5^3$:

$$\left(2^7\right)^7 \; > \; \left(5^3\right)^7$$

$$2^{49} \; > \; 5^{35}$$

Solution 3

Since $2^3 < 3^2$ and $2^7 > 5^3$, we have:

$$2^{63} = \left(2^3\right)^{21}$$

$$3^{42} = \left(3^2\right)^{21}$$

and

$$5^{27} = \left(5^3\right)^9$$

We can use the first two inequalities to find out right away that:

$$2^{63} < 3^{42}$$

We notice that:

$$2^{63} = \left(2^7\right)^9$$

which enables us to prove that:

$$2^{63} > 5^{27}$$

and, therefore, the order is: $5^{27} < 2^{63} < 3^{42}$

The answer is A.

Solution 4

Any of the fractions can be written as:

$$\frac{k}{50 + 2k}$$

with $k = 1, 2, \ldots, 99$.

Assume that among two consecutive terms in the sequence the first one is smaller than the second one:

$$
\begin{aligned}
\frac{k}{50 + 2k} &< \frac{k + 1}{50 + 2(k + 1)} \\
k(50 + 2(k + 1)) &< (k + 1)(50 + 2k) \\
50k + 2k(k + 1) &< 50k + 50 + 2k^2 + 2k \\
2k^2 + 2k &< 2k^2 + 2k + 50 \\
0 &< 50
\end{aligned}
$$

Since this inequality is valid for any value of k, the sequence is always increasing and the largest term is the last one.

Note that it is possible to multiply by the denominators because we know that they are positive numbers.

Solution 5

Compare 3^{445} with 4^{356}. Factor the exponents into primes to find a clue: $445 = 5 \cdot 89$ and $356 = 2^2 \cdot 89$. Since 89 is a common factor, we can compare:

$$3^{5 \cdot 89} = \left(3^5\right)^{89} \qquad \text{and} \qquad 4^{4 \cdot 89} = \left(4^4\right)^{89}$$

Since $3^5 = 243$ and $4^4 = 256$, we have that:

$$3^5 < 4^4$$

and

$$\frac{3^{444}}{4} < \frac{4^{355}}{3}$$

Solution 6

Subtract A from B:

$$
\begin{aligned}
B - A &= 1 \cdot 2 - 1 + 2 \cdot 3 - 2^2 + 3 \cdot 4 - 3^2 + \cdots 1000 \cdot 1001 - 1000^2 \\
&= 1 + 2 \cdot (3 - 2) + 3 \cdot (4 - 3) + \cdots + 1000 \cdot (1001 - 1000) \\
&= 1 + 2 + 3 + \cdots + 1000 \\
&= 500500
\end{aligned}
$$

B is larger than A.

Solution 7

We have to find out which number is larger 2^{160} or 5^{70}. For this, we look for simpler inequalities that we may use. We notice that, since $5^7 = 78125$ and $2^{16} = 65536$

$$2^{16} < 5^7$$

Therefore:

$$\left|2^{160} - 5^{70}\right| = 5^{70} - 2^{160}$$

and,

$$A = 25^{35} \div \left(5^{70} - 2^{160} + 16^{40}\right)$$

$$= 5^{70} \div \left(5^{70} - 2^{160} + 2^{160}\right)$$

$$= 1$$

Solution 8

Since:

$$2 < \sqrt{7} < 3$$

We can write that:

$$-3 < -\sqrt{7} \quad \rightarrow \quad 0 < 3 - \sqrt{7} < 1$$

Also:

$$0 < 3 - \sqrt{7} < 2n \quad \rightarrow \quad 0 < n$$

and

$$2n < 3 + \sqrt{7} < 6 \quad \rightarrow \quad n < 3$$

These inequalities imply that:

$$S = \{1, 2\}$$

and

$$|S| = 2$$

Solution 9

Since $7ab6$ is an even number, for it to be a perfect square it must be a multiple of powers of 4. Therefore, b can only be odd. The square root must end with 4 or 6. Since $8^2 = 64$ and $9^2 = 81$, we have the following candidates for perfect squares: 84^2 and 86^2. Indeed:

$$(80 + 4)^2 = 6400 + 2 \cdot 80 \cdot 4 + 16 = 6400 + 640 + 16 = 7056$$

and

$$(80 + 6)^2 = 6400 + 2 \cdot 80 \cdot 6 + 36 = 6400 + 960 + 36 = 7396$$

The range is $7396 - 7056 = 340$.

Solution 10

Use the triangle inequalities:

$$10 - x + x - 1 \; > \; x - 3$$

$$x - 3 + x - 1 \; > \; 10 - x$$

$$10 - x + x - 3 \; > \; x - 1$$

which simplify to:

$$12 \; > \; x$$

$$3x \; > \; 14$$

$$8 \; > \; x$$

If the side lengths are integer, x must also be an integer. The set of possible values for x is:

$$\{5, 6, 7\}$$

There are 3 possible values.

Solution 11

Since only one end of the interval is included in the set, the number of elements of the set A is:

$$2^{4672} - 2^{4671} = 2^{4671} (2 - 1) = 2^{4671}$$

Therefore, $|A| = 2^{4671}$.

Similarly, the number of elements in the set B is:

$$|B| = 5^{2002} - 5^{2001} = 5^{2001} (5 - 1) = 5^{2001} \cdot 4$$

Since $2^7 = 128$ and $5^3 = 125$ we have that:

$$2^7 > 5^3$$

and since $2001 = 3 \times 667$,

$$\left(2^7\right)^{667} > \left(5^3\right)^{667}$$

$$\left(2^7\right)^{667} \cdot 4 > \left(5^3\right)^{667} \cdot 4$$

$$2^{7 \cdot 667 + 2} > 5^{2001} \cdot 4$$

$$2^{4671} > 5^{2001} \cdot 4$$

Therefore, $|A| > |B|$.

Solution 12

Since the numbers are integers, so are their differences. Because 23 is prime and the numbers are different, the only possible factorization is:

$$a(a - b)(a - c) = (-1) \cdot (1) \cdot (-23)$$

Given the ordering, the only way we can assign values to a, b, and c is:

$$a = 1$$
$$b = 2$$
$$c = 24$$

Therefore, $c - b = 22$.

Solution 13

Since the digits could be placed in increasing order, we have to exclude the digit 0 from their possible values. The digits do not repeat, therefore there are 4! numbers that can be formed with the digits a, b, c, d in some order. Of these one number has digits in increasing order and one other number has digits in decreasing order.

The probability is:

$$P = \frac{4! - 2}{4!} = \frac{22}{24} = \frac{11}{12}$$

Solution 14

From the condition that $a + b < 6$ we have that the possible pairs (a, b) are:

$$(a, b) \in \{(1, 1), (1, 2), (1, 3), (1, 4), (2, 1), (2, 2), (2, 3), (3, 1), (3, 2), (4, 1)\}$$

Of these, only $(1, 4)$ does not satisfy: $b - a < 3$. Therefore, there are 9 pairs of positive integer numbers that satisfy both inequalities.

Solution 15

We have to find out how many 2-digit numbers satisfy:

$$ab - ba < 3 \cdot (a + b)$$

Write both the number and its reverse in expanded form:

$$
\begin{aligned}
10a + b - 10b - a &< 3a + 3b \\
9a - 9b &< 3a + 3b \\
3a - 3b &< a + b \\
2a &< 4b \\
a &< 2b
\end{aligned}
$$

Therefore, the larger digit has to be less than twice the smaller digit: $b < a < 2b$.

Since both the number and its reverse have to be 2-digit numbers, b must be a non-zero digit.

If $b = 1$ there is no possible value for a.
If $b = 2$, $2 < a < 4$ and $a = 3$ is a possible value.
If $b = 3$, $3 < a < 6$ and $a \in \{4, 5\}$.
If $b = 4$, $4 < a < 8$ and $a \in \{5, 6, 7\}$.
If $b = 5$, $5 < a < 10$ and $a \in \{6, 7, 8, 9\}$.
If $b = 6$, $a \in \{7, 8, 9\}$.
If $b = 7$, $a \in \{8, 9\}$.
If $b = 8$, $a = 9$ is the only possible value.

There are $1 + 2 + 3 + 4 + 3 + 2 + 1 = 16$ such 2-digit numbers. The total number of 2-digit numbers that have digits in decreasing order is:

$$\frac{9 \cdot 8}{2} = 36$$

The probability required is:

$$P = \frac{16}{36} = \frac{4}{9}$$

Solution 16

A number written in base 3 only with k digits of 2 has the form:

$$3^k - 1$$

Similarly, a number written in base 2 with 11 digits of 1 is equal to:

$$2^{11} - 1 = 2048 - 1 = 2047$$

We need the smallest N larger than 2047. Since $3^7 = 2187$:

$$2047 < 3^7 - 1$$

When written in base 3, $3^7 - 1$ is a number with 7 digits of 2: 2222222.

Solution 17

Notice that since $99 < 100$, we also have:

$$\frac{9}{10} < \frac{10}{11}$$

Applying a similar process to all the fractions in the product we get that:

$$\frac{9}{10} \cdot \frac{11}{12} \cdot \frac{13}{14} \cdots \frac{99}{100} < \frac{10}{11} \cdot \frac{12}{13} \cdot \frac{14}{15} \cdots \frac{100}{101}$$

Notice that:

$$\frac{9}{10} \cdot \frac{10}{11} = \frac{9}{11}$$

and, therefore:

$$\frac{9}{10} \cdot \frac{11}{12} \cdot \frac{13}{14} \cdots \frac{99}{100} \cdot \frac{10}{11} \cdot \frac{12}{13} \cdot \frac{14}{15} \cdots \frac{100}{101} = \frac{9}{101}$$

Denote by P, the product:

$$P = \frac{9}{10} \cdot \frac{11}{12} \cdot \frac{13}{14} \cdots \frac{99}{100}$$

and by Q, the product:

$$Q = \frac{10}{11} \cdot \frac{12}{13} \cdot \frac{14}{15} \cdots \frac{100}{101}$$

Then,

$$P \cdot Q = \frac{9}{101} < \frac{9}{100}$$

But $P < Q$, and:

$$P < \sqrt{\frac{9}{101}} < \sqrt{\frac{9}{100}} = \frac{3}{10}$$

Solution 18

First, let us attempt to find the number of digits of $4^{115} = 2^{230}$. Use:

$$2^{10} = 1024 > 10^3$$

Therefore, since 230 is a multiple of 10:

$$\left(2^{10}\right)^{23} > \left(10^3\right)^{23}$$

and 4^{115} has more than 69 digits. That is, it has at least 70 digits.

On the other hand, $5^3 = 125$ and $2^7 = 128$, so $5^3 < 2^7$:

$$
\begin{aligned}
5^7 &= 5^3 \cdot 5^3 \cdot 5 \\
&= (2^7 - 3)(2^7 - 3)(2^2 + 1) \\
&= (2^{14} - 6 \cdot 2^7 + 9)(2^2 + 1) \\
&= 2^{16} - 3 \cdot 2^{10} + 2^{14} - 3 \cdot 2^8 + 9(2^2 + 1) \\
&= 2^{16} - (4 - 1) \cdot 2^{10} + 2^{14} - (4 - 1) \cdot 2^8 + 9(2^2 + 1) \\
&= 2^{16} - 2^{12} + 2^{10} + 2^{14} - 2^{10} + 2^8 + 9(2^2 + 1) \\
&= 2^{16} + 2^{14} - 2^{12} + 2^8 + 9(2^2 + 1) > 2^{16}
\end{aligned}
$$

131

Since:

$$5^7 > 2^{16}$$

$$5^7 \cdot 2^7 > 2^{16} \cdot 2^7$$

$$10^7 > 2^{23}$$

and, therefore:

$$10^{70} > 2^{230}$$

As a result, 2^{230} must have at most 70 digits. Since we already know that it has at least 70 digits, then we have that it has exactly 70 digits.

$$5^{230} \cdot 2^{230} = 10^{230}$$

Since 10^{230} has 231 digits, we have that 5^{230} has exactly $231 - 70 = 161$ digits. Therefore, it has $161 - 70 = 91$ digits more than the number 4^{115}.

SOLUTIONS TO MISCELLANEOUS PRACTICE

Do not use a calculator for any of the problems!

Solution 1

Simplify the expression:

$$\frac{25^4 + 5^6}{169} \cdot \frac{65}{100000} = \frac{5^8 + 5^6}{13^2} \cdot \frac{5 \cdot 13}{2^5 \cdot 5^5}$$

$$= \frac{5^6(5^2 + 1)}{13^{\cancel{2}}} \cdot \frac{\cancel{5} \cdot \cancel{13}}{2^5 \cdot 5^4 \cdot \cancel{5}}$$

$$= \frac{5^6 \cdot 26}{13} \cdot \frac{1}{2^5 \cdot 5^4}$$

$$= \frac{5^{6-4} \cdot 2 \cdot \cancel{13}}{\cancel{13}} \cdot \frac{1}{2^5}$$

$$= \frac{5^2}{2^4}$$

$$= \frac{25}{16}$$

Since $\gcd(25, 16) = 1$, $m + n = 25 + 16 = 41$.

Solution 2

Use the identity $m^2 - n^2 = (m + n)(m - n)$:

$$
\begin{aligned}
29^2 - 21^2 - 16^2 - 12^2 &= (29 + 21)(29 - 21) - 16^2 - 12^2 \\
&= 50 \cdot 8 - 16^2 - 12^2 \\
&= 25 \cdot 16 - 16^2 - 12^2 \\
&= 5^2 \cdot 4^2 - 16^2 - 12^2 \\
&= (20 + 16)(20 - 16) - 12^2 \\
&= 36 \cdot 4 - 12^2 \\
&= 6^2 \cdot 2^2 - 12^2 \\
&= 12^2 - 12^2 \\
&= 0
\end{aligned}
$$

Solution 3

Use the identity $m^2 - n^2 = (m + n)(m - n)$:

$$
\begin{aligned}
65^2 - 60^2 - 24^2 &= (65 + 60)(65 - 60) - 24^2 \\
&= 125 \cdot 5 - 24^2 \\
&= 25^2 - 24^2 \\
&= (25 + 24)(25 - 24) \\
&= 49
\end{aligned}
$$

Therefore, $x = 7$.

Solution 4

Simplify the expression:

$$\frac{(3^{10} - 3^9)(3^8 - 3^7) \cdot \cdots \cdot (3^2 - 3)}{(3^2 - 2^3 + 1)^5 (2^5 - 5)^{15}} = \frac{\left(3^9(3-1)\right) \cdot \left(3^7(3-1)\right) \cdot \cdots \cdot 3(3-1)}{(10 - 8)^5 (32 - 5)^5}$$

$$= \frac{3^9 \cdot 3^7 \cdots 3^1 \cdot 2 \cdot 2 \cdots 2}{2^5 \cdot (27)^5}$$

$$= \frac{3^{9+7+5+3+1} \cdot 2^5}{2^5 \cdot (3)^{15}}$$

$$= \frac{3^{25}}{(3)^{15}}$$

$$= 3^{10}$$

Therefore, $k = 10$.

Solution 5

Use the factorization of $21 = 3 \cdot 7$:

$$\frac{21^2 \cdot 21^4 \cdot \cdots \cdot 21^{12}}{(2^3 - 1)^{42} (2^3 + 1)^{21}} = \frac{3^2 \cdot 7^2 \cdot 3^4 \cdot 7^4 \cdots 3^{12} \cdot 7^{12}}{7^{42} \cdot 3^{42}}$$

$$= \frac{3^{2+4+\cdots+12} \cdot 7^{2+4+\cdots+12}}{7^{42} \cdot 3^{42}}$$

$$= \frac{3^{2(1+2+\cdots+6)} \cdot 3^{2(1+2+\cdots+6)}}{7^{42} \cdot 3^{42}}$$

$$= \frac{3^{42} \cdot 7^{42}}{7^{42} \cdot 3^{42}}$$

$$= 1$$

Solution 6

Apply the definitions of the two means. The number is:

$$
\sqrt{\frac{20 + 60}{2} \cdot \frac{13 + 32}{2}} = \sqrt{\frac{80}{2} \cdot \frac{45}{2}}
$$

$$
= \sqrt{\frac{80 \cdot 45}{4}}
$$

$$
= \sqrt{\frac{16 \cdot 5 \cdot 9 \cdot 5}{2^2}}
$$

$$
= \frac{4 \cdot 5 \cdot 3}{2}
$$

$$
= 2 \cdot 5 \cdot 3
$$

$$
= 30
$$

Solution 7

Notice that the arithmetic mean is always larger than the geometric mean:

n	m	$AM(m, n)$	$GM(m, n)$
10	40	25	20
15	375	195	75
12	48	30	24
44	275	159.5	110

Solution 8

Since any two consecutive numbers are *coprime*, their gcf is equal to 1 and their lcm is equal to the product of the numbers. Therefore:

$$\text{lcm}(1, 2) \cdot \text{lcm}(3, 4) \cdot \cdots \cdot \text{lcm}(N, N + 1) = 1 \cdot 2 \cdot 3 \cdot 4 \cdots N \cdot (N + 1)$$

and

$$\frac{\text{lcm}(1, 2) \cdot \text{lcm}(3, 4) \cdot \cdots \cdot \text{lcm}(N, N + 1)}{(N + 1)!} = 1$$

Solution 9

Notice that all exponents are even, for any values of a and b that are positive integers:

$$ab^2 + a^2b = ab(a + b)$$

If any of a or b are even then the product is even. If they are both odd, then the sum is even and the product is also even.

Similarly a and $5a + 1$ have different parities: if one is odd the other is even and conversely. Their product is always even.

If b is odd $4a + b$ is odd but $b + 1$ is even, etc.

The sum is equal to 3. The correct answer is (C).

Solution 10

The sum is a perfect square:

$$1 + 3 + 5 + \cdots + k = p^2$$

Therefore, we have:

$$p = 2^2 \cdot 3 \cdot 7 = 4 \cdot 21 = 84$$

There are 84 terms in the sum. Since the common difference is 2:

$$k = 1 + 2(84 - 1) = 1 + 2 \times 83 = 167$$

Solution 11

$$
\begin{aligned}
G &= \left(2 + \frac{3}{2} + \frac{4}{3} + \cdots + \frac{2016}{2015}\right) + \left(4030 - \frac{1}{2} - \frac{2}{3} - \cdots - \frac{2014}{2015}\right) \\
&= \left(2 + 1 + \frac{1}{2} + 1 + \frac{1}{3} + \cdots + 1 + \frac{1}{2015}\right) + \left(4030 - \frac{1}{2} - 1 - \frac{1}{3} - \cdots - 1\frac{1}{2015}\right) \\
&= \left(2 + 1 + \cdots + 1 + \frac{1}{2} + \cdots + \frac{1}{2015}\right) + \left(4030 - 1 - \cdots - 1 - \frac{1}{2} - \cdots - \frac{1}{2015}\right) \\
&= \left(2 + 2015 + \frac{1}{2} + \cdots + \frac{1}{2015}\right) + \left(4030 - 2013 - \frac{1}{2} - \cdots - \frac{1}{2015}\right) \\
&= 2017 + 2017 \\
&= 4034
\end{aligned}
$$

Solution 12

The common difference is:

$$k \cdot n - n = n(k - 1)$$

Therefore, the third term is:

$$n + n(k - 1) \times 2 = n + 2nk - 2n = 2nk - n$$

and the fifth term is:

$$n + n(k - 1) \times 4 = n + 4nk - 4n = 4nk - 3n$$

For the fifth term to be divisible by the third term, the following fraction must reduce to an integer value:

$$\frac{4nk - 3n}{2nk - n}$$

Simplify by n:

$$\frac{n(4k - 3)}{n(2k - 1)} = \frac{4k - 3}{2k - 1}$$

This fraction is improper for any value of k. Separate the integers out of the fraction:

$$\frac{4k-3}{2k-1} = \frac{4k-2}{2k-1} - \frac{1}{2k-1} = 2 - \frac{1}{2k-1}$$

The result is an integer if and only if the fraction is an integer. For this, $2k-1$ must be a divisor of 1, i.e. it must be in the set $\{-1, 1\}$:

$$2k - 1 = 1 \quad \rightarrow k = 1$$

or

$$2k - 1 = -1 \quad \rightarrow k = 0$$

There are two values of k that satisfy the condition.

Solution 13

For a number that is a perfect square to be a perfect cube, it must be a perfect sixth power.

Notice that, since $2^{12} = 4096$, there are no perfect 12-powers or other powers that are multiples of 6.

Since $3^6 = 729$, only $2^6 = 64$ is a perfect sixth power smaller than 300.

The smallest combination of two different prime factors is $2^6 \cdot 3^6$, which is too large. Therefore, there is only one number that satisfies.

Solution 14

Two numbers differ by a multiple of 10 if they have the same last digit. The last digit of a perfect square is an element of the set:

$$\{0, 1, 4, 5, 6, 9\}$$

By the pigeonhole principle, if Damien has 7 cards, at least two of the cards will have numbers that have the same last digit.

Solution 15

Notice that 2015 and 2017 differ by 2:

$$\sqrt{4 + 2020\sqrt{1 + (2015 + 2) \cdot 2015}} = \sqrt{4 + 2020\sqrt{1 + 2 \cdot 2015 + 2015^2}}$$

$$= \sqrt{4 + 2020\sqrt{(1 + 2015)^2}}$$

$$= \sqrt{4 + 2020\sqrt{2016^2}}$$

$$= \sqrt{4 + 2020 \cdot 2016}$$

Notice that 2020 and 2016 differ by 4:

$$\sqrt{4 + 2020 \cdot 2016} = \sqrt{4 + (4 + 2016) \cdot 2016}$$

$$= \sqrt{4 + 4 \cdot 2016 + 2016^2}$$

$$= \sqrt{2^2 + 2 \cdot 2 \cdot 2016 + 2016^2}$$

$$= \sqrt{(2 + 2016)^2}$$

$$= \sqrt{2018^2}$$

$$= 2018$$

Notice that the result can also be -2018. However, earlier during the computation, we could not have used -2016 in the earlier step since it would have resulted in taking the square root of a negative number, an operation which cannot be performed in the realm of real numbers.

Solution 16

Notice that you can apply the identity $(a - b)(a + b) = a^2 - b^2$ twice:

$$
\begin{aligned}
P &= \sqrt{5 + \sqrt{3 + \sqrt{k}}} \cdot \sqrt{5 - \sqrt{3 + \sqrt{k}}} \cdot \sqrt{22 + \sqrt{k}} \\
&= \sqrt{25 - (3 + k)} \sqrt{22 + k} \\
&= \sqrt{22 - k} \sqrt{22 + k} \\
&= \sqrt{22^2 - k^2}
\end{aligned}
$$

For the result to be an integer $22^2 - k^2$ must be a perfect square:

$$
\begin{aligned}
22^2 - k^2 &= p^2 \\
k^2 + p^2 &= 22^2 \\
k^2 + p^2 &= 484
\end{aligned}
$$

Since perfect squares can only end in $\{0, 1, 4, 5, 6, 9\}$, for the sum of the two squares to end in 4, k and p must end in: $(0, 4)$ or $(5, 9)$. The only square smaller than 484 that ends in 0 is 100 and 384 is not a perfect square (since the sum of its digits is 15 which shows it is divisible by 3 but not by 9). Below 484, the only square that ends in 5 is 25 and $484 - 25 = 459 = 9 \cdot 51$, which is not a perfect square.

Therefore, there are no integer values of k that can make P an integer.

Solution 17

Simplify all the odd factors in the fraction on the left:

$$\frac{\not{1} \cdot 2 \cdot \not{3} \cdot 4 \cdots \cdots \not{111}}{\not{1} \cdot \not{3} \cdot \not{5} \cdots \cdots \not{111}} \cdot \frac{2^{-54}}{1 \cdot 2 \cdot 3 \cdots \cdots 55} =$$

$$\frac{2 \cdot 4 \cdot 6 \cdots \cdots 110}{1} \cdot \frac{2^{-54}}{1 \cdot 2 \cdot 3 \cdots \cdots 55} =$$

$$\frac{2 \cdot 2 \cdot 2 \cdots \cdot 2 \cdot \not{1} \cdot \not{2} \cdot \not{3} \cdots \not{55}}{1} \cdot \frac{2^{-54}}{\not{1} \cdot \not{2} \cdot \not{3} \cdots \cdots \not{55}} =$$

$$2^{55} \cdot 2^{-54} =$$

$$2$$

Solution 18

Notice that all the denominators are succesive triangular numbers:

$$\frac{49}{50} = \frac{2}{2 \cdot 3} + \frac{2}{3 \cdot 4} + \frac{2}{4 \cdot 5} + \cdots + \frac{2}{n(n+1)}$$

$$= 2\left(\frac{1}{2 \cdot 3} + \frac{1}{3 \cdot 4} + \frac{1}{4 \cdot 5} + \cdots + \frac{1}{n(n+1)}\right)$$

and now use the identity:

$$\frac{1}{n(n+1)} = \frac{1}{n} - \frac{1}{n+1}$$

142

to telescope the sum:

$$\frac{49}{50} = 2\left(\frac{1}{2} - \frac{1}{3} + \frac{1}{3} - \frac{1}{4} + \cdots + \frac{1}{n} - \frac{1}{(n+1)}\right)$$

$$= 2\left(\frac{1}{2} - \frac{1}{(n+1)}\right)$$

$$= 2\left(\frac{n+1-2}{2(n+1)}\right)$$

$$= 2\frac{n-1}{2(n+1)}$$

$$= \frac{n-1}{n+1}$$

Solve the equation:

$$\frac{49}{50} = \frac{n-1}{n+1}$$

$$49(n+1) = 50(n-1)$$

$$49n + 49 = 50n - 50$$

$$99 = n$$

Solution 19

Square both sides:

$$\left(x + \frac{1}{x}\right)^2 = 49$$

$$x^2 + 2 + \frac{1}{x^2} = 49$$

$$x^2 + \frac{1}{x^2} = 47$$

And now calculate:

$$\left(x - \frac{1}{x}\right)^2 \quad = \quad x^2 - 2 + \frac{1}{x^2}$$

$$= 47 - 2$$

$$= \quad 45$$

Therefore, $x - \frac{1}{x} = \sqrt{45} = 3\sqrt{5}$.

Math Challenges for Gifted Students

Practice Tests in Math Kangaroo Style for Students in Grades 1 – 2
Practice Tests in Math Kangaroo Style for Students in Grades 3 – 4
Practice Tests in Math Kangaroo Style for Students in Grades 5 – 6

Weekly Math Club Materials for Students in Grades 1 – 2

Competitive Mathematics Series for Gifted Students

Practice Counting (ages 7 to 9)
Practice Logic and Observation (ages 7 to 9)
Practice Arithmetic (ages 7 to 9)
Practice Operations (ages 7 to 9)

Practice Word Problems (ages 9 to 11)
Practice Combinatorics (ages 9 to 11)
Practice Arithmetic(ages 9 to 11)
Practice Operations (ages 9 to 11)

Practice Word Problems (ages 11 to 13)
Practice Combinatorics and Probability (ages 11 to 13)
Practice Arithmetic and Number Theory (ages 11 to 13)
Practice Algebra and Operations (ages 11 to 13)
Practice Geometry (ages 11 to 13)

Self-help:

Parents' Guide to Competitive Mathematics

Coming Soon:

Weekly Math Club Materials for Students in Grades 3 − 4
Weekly Math Club Materials for Students in Grades 5 − 6

Practice Word Problems (ages 12 to 16)
Practice Algebra and Operations (ages 12 to 16)
Practice Geometry (ages 12 to 16)
Practice Number Theory (ages 12 to 16)
Practice Combinatorics and Probability (ages 12 to 16)

This is a series of practice books. With the exception of a few reminders, there are no theoretical explanations. Online problem solving lessons are available at www.goodsofthemind.com. If you found this booklet useful, you will enjoy the live problem solving lessons.

For additional problem solving material, look up our series *Math Challenges for Gifted Students* which includes practice tests in Math Kangaroo style and weekly math club materials.

Made in United States
North Haven, CT
29 October 2022

26088420R00083